BOOK OF SCIENCE STUFF

BOOK OF
Science
Stuff

Joe Rhatigan

Illustrated by Jeff Albrecht Studios

imagine!
Publishing
New York
www.imaginebks.com

Text and art copyright © 2010 by Imagine Publishing, Inc.

Published by Imagine Publishing, Inc.

Library of Congress Cataloging-in-Publication Data
Rhatigan, Joe.
Book of science stuff / Joe Rhatigan.
p. cm.
ISBN 978-1-936140-18-3 (hardcover : alk. paper)
1. Science—Popular works. I. Title.
Q162.R43 2010
500—dc22
2010006765

Designed by Marc Cheshire
Printed in China
1 3 5 7 9 10 8 6 4 2

To Charlie and Jeremy . . . thank you!

Contents

Kitchy, kitchy, koo, koo, you big ape.

Introduction

In a 2009 study published by *Current Biology*, a team of scientists asserted that laughter is not a unique human trait. How did these scientists (whose parents probably worked two jobs each to pay for their educations) figure out that other animals laugh? They tickled baby apes.

The scientists concluded that orangutans, gorillas, chimpanzees, and bonobos are, indeed, ticklish. Now, this study has serious implications, and several scientists praised the findings. But that's not what this book is about. *Science Stuff* will not bore you with such things as "important findings." No, this book celebrates the offbeat moments, the ridiculous findings, the screw-ups, the goofballs, and the often-clueless humans who practice and experience all the fields of science. Because really, what's more interesting, the fact that scientists are closer to understanding our links with our ancestral cousins, or that a highly respected team of scientists (it took more than one!) tickled apes in the name of science? If you answered the latter, this book is for you.

Sure, science is serious business. The word itself reeks of chemicals, musty textbooks, lectures that threaten to dislocate your jaw from serial yawning, and nerds. But behind the lab coats and Coke-bottle glasses lies a very important fact. If you can wade through all those pesky theories, hypotheses, and world-altering discoveries, you'll find that the field of science reads more like a series of punch lines than a research paper riddled with footnotes and references. That's because no matter how objective science strives to be, scientists are human—the least objective life forms in the universe (to our knowledge). And there's the rub. We may not be the only species that can laugh, but as far as we know, we're the primary source of all the humor on this planet. So, without taking anything away from all that science and its practitioners do for the advancement of society, we present our own humble findings. We spent hours researching, studying, culling, and goofing off, and the results are this subjective, stilted view of the majestic endeavor of science as practiced and experienced by our all too fallible race.

You can read *Science Stuff* straight through, read it from back to front, or keep it in the bathroom for multi-tasking purposes. Whatever you do, read it to your chimp. He can probably use a good laugh.

The Scientific Method in Action

IMAGINE you're sitting under an apple tree when suddenly an apple falls on your head. If you're a normal person, you may rub your head and say "Ouch," throw the apple in anger, or even eat it. If, however, you're a more curious creature, you may do all those things and then start wondering why the apple fell from the tree in the first place or why apples and rocks and falling pianos don't stay in the air. If you're a scientist, you stop what you're doing and devote the rest of your life (often ignoring loved ones and your own personal hygiene) testing why that apple fell from the tree. The scientist doesn't just make assumptions about the apple. She or he follows a method of discovering knowledge that includes making predictions, testing these predictions empirically, and then developing theories that best explain the results. This is known as the *scientific method*, and scientists have been using it for a thousand years to ascertain why it is men always leave the toilet seat up or whether fish get drunk when exposed to alcohol. The scientific method has several guidelines, none of which state that "just because you can, doesn't mean you should."

BLUNDEROLOGY

Perhaps Einstein said it best: "Two things are infinite: the universe and human stupidity, and I'm not sure about the universe." As we've already proven, scientists are human. Hence, scientists are sometimes stupid.

The Tooth Scary

Aristotle, the ancient Greek philosopher and scientist, is considered one of the founding figures of Western philosophy whose work in the fields of science continues to influence scientists more than 1,500 years after his death. As fond of empirical study as he was, this didn't stop Aristotle from stating that women had fewer teeth than men . . . without bothering to check.

Sheep for Brains

Imagine UK government officials' consternation at having to inform the public that bovine spongiform encephalopathy (mad cow disease) had jumped the species barrier and infected sheep. The Institute for Animal Health in Edinburgh spent more than five years and nearly $400,000 of taxpayers' money studying sheep brains and was certain of its dire findings—meaning certain death for thousands of sheep. Unfortunately for the scientists (and fortunately for the sheep), because of a labeling problem and some other unexplained mix-ups, the scientists in charge of the study had been studying the wrong brains the whole time. Instead of sheep brains, they had been working with infected cow brains.

Shake, Rattle, and Die

Dr. Vladimir Gavreau was a French robotics researcher who developed remote-controlled devices for factories and the military at the height of the Cold War. His team worked in a large concrete building, and periodically they would all experience a "disconcerting nausea." The source of the sickness was found to be an improperly installed motor-driven ventilator, which was creating an infrasonic resonance (low-frequency sound no one could hear). Being in the weapons business, Gavreau believed he had found a new way to beat the bad guys. His team went about creating a sonic weapon, a giant infrasound organ with pipes six feet in diameter and up to seventy-five feet long. The first time Gavreau turned on the device, the building they were outside of was nearly destroyed and the researchers were gripped in what they later called "an envelope of death." Though the organ was only on for a few seconds (one of the researchers managed to turn it off), the scientists were extremely sick for the next twenty-four hours, and their eyesight was affected for days afterward. According to one report, their body cavities absorbed the acoustic energy and would have been torn to pieces had the power not been turned off quickly.

The experiment was considered a success in that they had indeed created an expedient way of eliminating an enemy, even as they nearly eliminated themselves.

So, what ever happened to Dr. Gavreau's sonic death ray? Well, since Gavreau was never able to create a suitable defense against his own creation, anyone who used it would die along with their enemy.

You Get What You Pay For

The Mars Climate Orbiter was one of a series of missions in a long-term program of Mars exploration, known as the Mars Surveyor Program. In September 1999, the craft approached Mars and then disappeared. At first, politicians and some scientists blamed NASA's new credo, "better, faster, cheaper," for the $125 million failure. But a week later, NASA's scientists figured out what happened, and they weren't exactly shouting it from the rooftops. It seems that even though

NASA has used metric units to guide its spacecrafts for years, Lockheed Martin, the company they hired to engineer the craft, used English units for its thrust data. This caused the craft's thrusters to plunge the Orbiter to its doom. Soon after the incident, Noel W. Hinners, vice president for flight systems at Lockheed Martin Aeronautics and master of the obvious, said, "We should have converted."

Curiosity Killed the Elephant

In an experiment that could have only happened in the 1960s, scientists decided to inject Tusko the elephant with a new drug that had been recently introduced to the world: the hallucinogenic LSD. Scientists wanted to see if the drug could induce musth (the state of elephant temporary madness that makes males aggressive). The unstated mission of this experiment: to see what would happen.

Not sure how much LSD to inject into an elephant, the team decided to use nearly 300 milligrams, which is about

3,000 times the typical human dosage. Soon after the injection, Tusko trumpeted furiously, got all wobbly, keeled over, and died. The article the scientists published a few months later noted that "It appears that the elephant is highly sensitive to the effects of LSD." The scientists, in their own defense, said they had taken the drug the day before the experiment and they were fine.

MONKEYING AROUND

Monkeys and other primates are our closest living cousins. Too bad for them, as that makes them perfect for scientists to study. Don't worry; no animals were hurt while creating this chapter.

Monkeys with Mouses

The old saying goes that if you give an infinite number of monkeys computers, they would sooner or later reproduce the works of Shakespeare. In 2003, Plymouth University researchers in the United Kingdom received a grant of nearly $4,000 to see what would happen if they gave six Sulawesi crested macaque monkeys at the Paignton Zoo one computer. Did they write any words from Hamlet? Did they write any words at all? Well, let's just say that between throwing feces at the computer and peeing on it, the monkeys managed to type out five pages of text—mostly the letter "s." The researchers were slightly encouraged by the fact that by the end of the experiment, the monkeys added the letters "a," "j," "l," and "m" to their repertoire.

Can You Say "Sexual Discrimination"?

Koko, known as the talking gorilla, communicates using sign language. She can sign more than 1,000 words as well as understand 2,000 more. According to a 2005 lawsuit, one of the things Koko likes to ask is for human females to show her their nipples. The suit claims that the president of the Gorilla Foundation in Woodside, California, who works with Koko, demanded that two employees do as Koko asked in order to "bond" with the gorilla. The employees refused to disrobe for Koko and were eventually let go. The lawsuit was settled out of court and neither side has disclosed the terms of the settlement. Koko isn't available for comment.

Pay Per View

A study that appeared in *Current Biology* in 2005 reported that male rhesus macaque monkeys willingly paid (in the form extra juice) for the privilege of staring at photographs of female monkey bottoms. They also gave up the juice for photos of high-ranking primates. Conversely, the scientists had to pay the monkeys to look at images of lower-ranking monkeys. Before you get any ideas about which not-safe-for-work website funded this research, the scientists were attempting to establish how primates acquire visual information about others in order to better understand the social machinery in the brain. They hope the information gathered

I'll drink to that!

here will one day help people suffering from autism.

Mom?

University of Wisconsin psychologist Harry Harlow conducted a series of experiments in the 1950s and '60s that tested motherly love in monkeys. In one experiment, he raised infant macaques in isolation chambers for two years. Let's just say things weren't good for these monkeys afterward. Harlow's best-known experiment involved removing rhesus monkeys from their mothers and offering them a choice of a wire mannequin monkey covered in soft terrycloth or a wire mother with no covering. In one cage, the soft mom didn't provide any food. Ugly wire mom did. In another cage, it was the opposite. The monkeys in both cages preferred soft mom. Monkeys raised with only an ugly wire mom, however, had trouble digesting their food and were often visually distressed.

Not surprisingly, many people thought these experiments

THE SCIENTIFIC METHOD IN ACTION

despicable. In fact, Harlow's psychology department's address, 600 N. Park, soon became known as GOON Park. As awful as these studies were for the monkeys, Harlow successfully showed that intimate contact is important for monkey children and, hence, human children, and his studies have helped psychologists work with abused children. His experiments also inadvertently led to the birth of the animal liberation movement.

ANIMAL VICE

Have you ever seen a squirrel pick up an acorn with both front paws and start gnawing on it? Isn't it cute how nearly human that squirrel looks in that instant? The animals in these next stories also seem nearly human. Cute? Not so much.

Penguins of the Night

Scientists have found ladies of the night even at the South

Pole. They wear black-and-white tuxedoes and accept small stones for payment.

In the late 1990s, Dr. Fiona Hunter, a researcher in the Zoology Department at Cambridge University, was studying the mating patterns of Adelie penguins on Ross Island, about 800 miles from the South Pole, when she noticed some interesting activity at the fringes of a colony.

First, some background: Stones are essential for penguins' nests. Penguin couples use them to build a platform that will keep their egg high and dry. Females with partners often leave their nest to find these valuable commodities, and they will steal stones from each other and fight over them.

According to Hunter, some females have come up with another strategy—they mate with single male penguins for exchange for stones from their nests. Once the mating is over, the female picks up her stone for payment and carries it back to her unsuspecting "official" mate. Hunter also observed some of the females returning to the single male's nest

Is that a rock in your pocket ... or are you just glad to see me?

and taking more stones without having to do anything. One female managed to gather sixty-two rocks in this fashion.

Hunter doesn't know why some females do this, but theorizes that they could be testing out future mates in case their current one dies before the next mating period. As for the males, they appear to have only one motive for their behavior, and it's the same one that drives human males to buy sports cars and hair products.

Getting in the Mood

According to a 2006 AP report, scientists in China have come up with an ingenious method of getting endangered and reluctant pandas to procreate: show them "racy" DVDs of pandas mating. Zhang Zhihe, a leading expert in panda porn, said that it worked and that births were up more than 50% compared to the previous year. Scientists haven't been able to get this to work outside of China yet and have resorted to artificial insemination.

Party Animal

Although there are several animals known to drink alcohol (robins have been recorded dropping off their perch after eating grapes that have fermented on the vine), researchers have found that the Malaysian pentailed shrew lives on a diet of nearly 100% beer, which is equivalent to a human drinking nine glasses of wine a day. They drink the fermented nectar of the bertam palm plant, which is about 3.8% alcohol. The most boggling aspect of this discovery is that these animals never get drunk. One researcher figures that if they can find out how these animals can cope with the alcohol, they can help humans deal with alcohol poisoning.

Drinks Like a Fish

Do animals get drunk? You probably don't have to do a scientific study to guess the answer to this question; however, there are several scientists out there who just had to see for themselves. And guess what? The answer is "Yes"!

Damn, now that's *firewater!*

According to the book *Drunken Goldfish & Other Irrelevant Scientific Research* by William Hartston, mice and rats made alcoholic by scientists have the following symptoms during withdrawal from alcohol: tail stiffening, teeth chattering, hair standing on end, wet shakes, sudden propulsion, whole body rigidity, and more. Cats, dogs, and other animals showed many of the same symptoms. But it's the goldfish of the title of this book that got most of the author's attention.

Goldfish immersed in 3.1% alcohol "fall over when drunk." They lose their righting reflex and they tend to forget things. In fact, if you teach a goldfish something when it's sober, it will probably forget it when drunk; but if you teach it something when it is drunk, it will forget it when placed in nonalcoholic water. Amazingly, the same results apply to humans.

The Morning After

Speaking of alcohol, scientists from Aristotle to Darwin and beyond have noticed that some animals drink alcohol without humans' help. Bees and elephants being but two examples. Aristotle once noted that people often trapped monkeys by leaving out jars of wine for them to drink. Once drunk, they would pass out. It was Darwin, however, who studied what happened the morning after. In his book *The Descent of Man*, he noticed some similarities between man and monkey the morning after a rousing good evening:

"On the following morning they (the monkeys) were very cross and dismal; they held their aching heads with both hands, and wore a most pitiable expression: when beer or wine was offered them, they turned away with disgust. . . ."

MILITARY INTELLIGENCE

Science has often been called into service in times of desperate need and wars (both cold and hot). Sometimes science then calls upon members of the wild kingdom . . . or the sun or even the moon to assist in gaining some sort of advantage over an adversary. And whenever you have competing world powers striving for any potential edge that would prove they are the smartest, most powerful, most destructive nation in the world, it's a good time to be a mad scientist.

Me-ow!

The first-known animal recruited to spy on an enemy was nicknamed Acoustic Kitty. The CIA (who else!) launched this top-secret project (of course) in the 1960s (naturally). It involved using house cats on spy missions. First a microphone and battery were implanted into a cat—along with an antenna in its tail. Then the cat was trained not to run after birds and mice. The procedure, along with the training, is said to have cost more than $60 million. The first cat mission was to listen in on some Soviets in Washington, DC. The cat was released near the men and almost immediately run over by a taxi and killed.

Navy Seals and Other Armed Farces

Scientists at a U.S. weapons lab in New Mexico reported in 2009 that they have trained honeybees to sniff out explosives. By exposing the insects to the odors of several types of explosives followed by a sugar-water "reward," the research-

The CIA's newest spy didn't know what hit her.

ers believe the bees will be able to detect dynamite, C-4, and other explosives, and alert their handlers by sticking out their proboscis.

Meanwhile, the U.S. Navy has trained dolphins to find

and report the existence of underwater mines. They've also trained dolphins and their sea-faring friends, the sea lions, to detect "water-borne intruders" near piers and ships.

Not to be outdone, scientists at the Robot Engineering Technology Research Center of East China's Shandong University stated in 2009 that they can now control a pigeon's movements. Through the use of electrodes, scientists stimulate different parts of the bird's brain via computer to force the bird to follow their commands. So far the birds can fly right, left, up, and down.

Finally, as reported in *Science World* in 2003, Marines in Kuwait bought more than forty chickens as a low-tech chemical detection system. The logic goes like this: If there's a chemical attack, the chickens die first, alerting the Marines to put on their gas masks.

Robo Bugs

Forget low-tech chickens. It's time to talk about cyborg bugs!

Yes, scientists at the U.S. Pentagon are hard at work growing insects with electronics inside them. These moths and beetles can be controlled remotely, and military uses for these insects of doom are countless. They could conduct reconnaissance missions, carry little weapons with a big bang, and even be armed with bio weapons. The Defense Advanced Research Projects Agency (DARPA), whose mission is "to prevent technological surprise for us and to create technological surprise for our adversaries," is excited about the potential possibilities of these "unmanned air vehicles." But before you start getting paranoid about the flies buzzing around you as you check out your library books, the robo bugs aren't quite operational outside of lab conditions . . . yet.

The Sun Gun

Every burgeoning world power would love to be able to destroy their enemy from space. During World War II, Nazi physicists, gathered at a large facility in Hillersleben,

Germany, secretly worked on several weapons. One of their grandest (and most sinister) projects was known as the *sonnengewehr*, or sun gun. The idea was to shoot a nearly two-mile-square concave mirror into space and use it to reflect sunlight onto the enemy. No, the Nazis weren't trying to tan their enemies. The sun gun would burn cities, boil reservoirs, and pretty much melt people. Like the bratty kid with a magnifying glass leaning over an anthill, the Nazis believed that their mirror of doom could vanquish their enemies. Unfortunately for them, they were vanquished before they could get their mirror out of the planning stages.

The Ice Boat

The Germans weren't the only ones gathering their mad scientists to come up with wacky weapons. Wanting to create an unsinkable ship to use against German U-boats in the mid-Atlantic, the Brits came up with one possible solution: Create an aircraft carrier out of . . . ice! (Hey, you can't sink ice!) Known as Project Habbakuk, this ship would be 2,000 feet long, with walls forty feet thick. Scientists and engineers used a material called Pykrete, which is a mixture of ice and wood pulp, to build a sixty-foot model in Canada. It worked, but it was slow, and engineers realized that a full-scale ship would require more than 8,000 people and at least a year to build. Project Habbakuk was eventually scrapped because of its impracticality and its extreme cost. It took more than three years for the model to melt.

One Big Nuclear Explosion for Mankind

Before the United States realized the dream of sending a man to the moon, certain scientists and Air Force officials had another plan to display U.S. superiority to the Soviets: Nuke the moon!

Dr. Leonard Reiffel, the only person talking about this top-secret plan, claims that they would explode a nuclear weapon at the lunar terminator, which is the dividing line between the dark and light side of the moon. This would make the explosion visible from Earth—ostensibly striking fear into the enemy (and everyone else, as well). The project was called, innocently enough, Project A119 or A Study of Lunar Research Flights.

YOU GOT A GRANT TO STUDY THAT?

For every Einstein out there changing the world with his or her theories, there are at least 100 scientists who think it a good use of their time to figure out whether or not the five-second rule is valid or if cows are happier if you give them names.

The Ig Nobel Prizes

Created as a parody of the Nobel Prizes, the Ig Nobel Prizes are given each year to scientists, inventors, and others who in the words of the magazine sponsoring the awards, the *Annals of Improbable Research*, "first make people laugh, and then make them think." Each year, ten prizes are awarded to people whose achievements "cannot or should not be reproduced." Below are some of the proud winners of an Ig.

Robert Matthews of Aston University in England won the Ig Nobel Physics Prize for his work attempting to discover why buttered toast tends to fall on its buttered side. His study, called "Tumbling Toast, Murphy's Law and the Fundamental Constants," was published in the *European Journal of Physics* in 1995.

The 2009 Ig Nobel Medicine Prize was awarded to Dr. Donald Unger of Thousand Oaks, California. Every day for sixty years, Dr. Unger cracked the knuckles on his left hand, but never the knuckles on his right hand, in order to see whether or not cracking your knuckles causes arthritis. According to the patient doctor, it doesn't. He told the British newspaper the *Guardian*, "After sixty years, I looked at my knuckles and there's not the slightest sign of arthritis. I looked up to the heavens and said: 'Mother, you were wrong, you were wrong, you were wrong.'"

Buck Weimer was awarded the Ig Nobel Biology Prize in 2001 for his invention, which is described as "Protective Underwear with Malodorous Flatus Filter." In other words, airtight underwear with a filter that removes gas before it escapes. He invented it for his wife who suffers from Crohn's disease, but he has been kind enough to offer his invention online at www.under-tec.com. As his slogan says, "Wear them for the ones you love."

Speaking of underwear, Takeshi Makino won the 1999 Chemistry award for his involvement with S-Check, an infidelity detection spray that wives can apply to their husband's underwear.

The 2001 Ig Nobel Astrophysics award was presented to Dr. Jack Van Impe and Rexella Van Impe for their discovery that black holes fulfill all the technical requirements for the location of hell.

The 2002 Peace Prize was presented to Keita Sato, Dr. Matsumi Suzuki, and Dr. Norio Kogure for promoting peace and harmony between species by inventing Bow-Lingual, a computer-based, automatic, dog-to-human, language-translation device.

Finally, the 1998 award in Literature was presented to Dr. Mara Sidoli of Washington, D.C., for her report, "Farting as a Defense against Unspeakable Dread."

The Fleecing of America

Speaking of dubious awards, William Proxmire, a U.S. Senator for Wisconsin from 1957 to 1989, was famous for issuing Golden Fleece of the Month awards, which identified "wasteful, ironic, or ridiculous uses of the taxpayers' money." Although many of his awards went to pork projects, quite a few went to scientific endeavors. The awards quickly catapulted Proxmire onto the national scene, even as scientists cried foul, saying that he recklessly attacked real research for his own gain. What follows are some Fleece Award highlights and lowlights. Legitimate science or wasteful spending? You decide.

- Proxmire's first Fleece award was given to the National Science Foundation in 1975 for funding an $84,000 study that attempted to ascertain why people fell in love. According to one of the scientists involved, it was actually an effort to "determine the extent to which the major cognitive and emotional theories could tell us something about the nature of passionate love and sexual desire." Does that sound like the same thing to you?

- In 1976, Dr. Robert Baron was awarded the Golden Fleece for his study on a temperature-aggression link, especially as riots swept through the United States during the summers in the late '60s and '70s. Proxmire renamed this study The Effects of Scantily Clad Women on the Behavior of Chicago's Male Drivers.

- Proxmire criticized Ronald Hutchinson for studying why monkeys clench their jaws. Hutchinson's research sought to define objective measures of aggression by evaluating animal behavior patterns. Hutchinson sued and the case was eventually settled out of court.

- The Federal Aviation Administration was given the award for spending $57,800 on a study of the physical measurements of 432 airline stewardesses, paying special attention to the "length of the buttocks" and how their knees were arranged when they were seated.

Ma'am, can you hold still for a minute?

- The Justice Department won for conducting a study on why prisoners wanted to get out of jail.

- What did the National Institute of Mental Health have to do to get an award? They funded a study of a Peruvian brothel. The researchers famously told the *New York Times* that they had to make repeated visits to ensure accuracy.

- The National Institute on Drug Abuse funded a project by psychologist Harris Rubin on developing objective evidence concerning marihuana's effect on sexual arousal by exposing groups of male pot-smokers to pornographic films. Senator John McClellan got into the act, opining, "I am firmly convinced we can do without this combination of red ink, blue movies, and Acapulco gold."

Drip . . . Drop

Calling this a long-term experiment is an understatement. In 1927, Professor Thomas Parnell of the University of Queensland in Brisbane, Australia, wanted to show his students that some substances that appear to be solid are actually high-viscosity fluids. He poured a heated sample of tar pitch (pitch is the name of any highly viscous liquid) into a sealed funnel and waited for it to drip. Parnell's students were rewarded with the first drip . . . in 1938. Experimenters have calculated that drops fall over a period of about ten years, giving this pitch a viscosity that's 230 billion times that of water.

The Five-Second Rule Debunked

The next time you drop your sandwich on the floor, you may want to think twice before invoking the five-second rule—you know, the rule that states that as long as you pick up the sandwich (or cookie, or waffle piece) within five seconds, bacteria won't have time to attach to your food item. In 2003, a high school student named Jillian Clarke dropped Gummi Bears and fudge-striped cookies onto ceramic tiles treated with E. coli and picked them up before five seconds passed. The results? Let's just say you had better keep your floors spotless or work on your hand-to-mouth coordination.

No Tipping Please

Liz Shapiro, an anthropologist at the University of Texas, decided to use her scientific research skills to answer an age-old question: Why don't pregnant woman tip over? She and two researchers at Harvard studied nineteen pregnant woman and decided that they don't tip over because they lean back . . . and their lower back curve extends across three vertebrae, which is one vertebrate longer than the curve of men's lower backs. No word on why men with large beer bellies don't tip over. Oh, wait, they often do!

Happier Than a Pig in Air-Conditioning

Scientists at the Agricultural Research Council in Cambridge, England, performed a five-year study attempting to ascertain what makes pigs happy. The pigs were housed in an air-conditioned sound-proof room. Needless to say, the pigs were pretty happy.

Got Name?

A study published in the journal *Anthrozoös* reports that dairy cows that have been given names produce more milk. Catherine Douglas, an animal behaviorist behind the research, states that this finding "reflects the humans' attitudes toward the cows, and therefore how they behave around them." In other words, cows with names are treated more humanely and well-cared-for cows produce more milk.

Frosty the Microscopic Snowman

The world was rocked in late 2009 (during a slow news cycle) when scientists at the National Physical Laboratory in West London created the world's smallest snowman. Measuring just .01mm across, this as-of-yet unnamed fella was created from two tiny beads that are usually used to help calibrate electron microscope lenses. The scientists used tools designed for manipulating nanoparticles to construct the snowman, and the face was "drawn" by an ion beam. And in a too-cute-for-words moment, they used a tiny blob of platinum for the nose.

Boo Boo Science

Your mom was right—ripping off bandages quickly is less painful than the slow approach. Dr. Carl O'Kane of James Cook University in Australia applied sticky bandages to sixty-five medical students and then either removed them slowly or quickly. The students ranked their pain from zero to ten. The scores proved that fast is best. O'Kane explained that his research found that pain is more of a psychological issue. In fact, pain isn't just what you perceive, but what you think you will perceive. So, if Mom told you pulling the bandage off slowly would hurt, it did.

Is Kansas as Flat as a Pancake?

Thanks to science, we can now answer that question definitively. Mark Fonstad and William Pugatch of Texas State University and Brandon Vogt of Arizona State University used topographic data of Kansas from a digital scale model and then purchased a pancake from the International House of Pancakes. If perfect flatness is a value of 1, they calculated the flatness of a pancake to be .957. Kansas, however, came out flatter at a whopping .997.

PSYCHOLOGICALLY DISTURBED

As badly as many scientists have treated animals, it's pretty amazing to see how badly they've also treated humans.

On Being Sane in Insane Places

The title of this section was the name of the paper psychologist David Rosenhan published in *Science* in 1973. And the study in the article, indeed, took sane people and placed them in psychiatric hospitals to see if they could get released without any external help. How difficult could it be, anyway? After all, they were sane . . . weren't they?

Rosenhan and eleven volunteers each went to a different hospital in various locations within the United States and simulated auditory hallucinations (they pretended to hear voices). All were admitted, and once in the hospital, they proceeded to act normally and say they felt just fine. The staff at each hospital believed every one of these pseudo patients suffered from a mental illness—not one doctor or nurse caught on, even though more than 20% of the real patients believed they were fakes. Each pseudo patient took notes on the doctors' and nurses' actions, and one nurse called this "writing behavior" and labeled it pathological. Ultimately, each imposter did get released, although a lawyer had to be hired, and in some cases, it took nearly two months to get out. Each imposter also had to admit to having a mental illness and promise to take antipsychotic drugs. In a second experiment, Rosenhan told several hospitals he was going to send imposters to try to gain entry. The hospitals falsely identified many real patients as sane. How did Rosenhan figure this out? He never sent any imposters to the hospitals. Rosenhan concluded his study thusly: "It is clear that we cannot distinguish the sane from the insane in psychiatric hospitals." You don't say. . . .

Doing Time

Back in 1971, Stanford University psychology professor Philip Zimbardo had no idea that his experiment designed to study the effects of becoming either a prisoner or a prison guard would appear in nearly every psychology textbook and become the basis for novels, documentaries, songs, and even a feature-length film, to be released forty years after it was all over.

It was a simple two-week experiment. He chose twenty-four "normal" male students and randomly assigned them the role of either prisoner or guard. This prison simulation took place in the basement of the psychology building. A research assistant was the warden and Zimbardo, the superintendent. The guards were told not to physically harm anyone and were given wooden batons and uniforms. The prisoners were arrested at their homes and taken to jail.

The experiment quickly grew out of control. On day two, the prisoners rioted. The guards quickly squashed it by squirting the prisoners with fire extinguishers. After a rumor spread about a possible breakout attempt, the guards dismantled the jail and moved the inmates to a different location. Angry at the prisoners' behavior, the guards raised things up a notch. They forced prisoners to count off repeatedly, made them exercise for extremely long periods of time, refused to let them use the bathroom (or to empty the pail they were forced to use), made them sleep on the concrete floor naked, and even had them clean toilet bowls . . . with their hands. Remember, this was an experiment. This fact seemingly eluded the "guards," the "prisoners," the several onlookers, and even the "superintendent" Dr. Zimbardo, who admitted that he, along with the other participants, had internalized their roles.

It wasn't until Zimbardo's girlfriend (and future wife) visited and expressed her horror at what was going on did Zimbardo end the fiasco nine days early. Researchers claim that up to one-third of the guards had shown genuine sadistic tendencies, and many of the prisoners showed signs of trauma. (Two prisoners even had to be removed from the study early.)

What did it all mean? Are humans too susceptible to the power of authority? Is anyone capable of doing unthinkable things? Was there an ethics code review board in 1971?

Hey, aren't you in my sociology class?

These questions and many more that were raised by this experiment are still being discussed today.

Totally Shocking

Yale psychologist and (bizarrely enough) former classmate of Philip Zimbardo (see previous story) Stanley Milgram wanted to measure people's willingness to follow authority. He ran his experiment soon after the Nuremberg trial of German war criminal Adolf Eichmann started, and he was curious to find out if there was "a mutual sense of morality among those involved" when carrying out the atrocities in the concentration camps.

Milgram set up a situation where a test subject was told he was a teacher who was to give a memory test to another subject—the learner—who was in another room. (This person was actually an actor.) The teacher was then instructed to give the learner a memory test, and whenever he gave an incorrect answer, the teacher was to press a button that shocked the learner. An experimenter (also an actor) was in the room with the teacher to give verbal encouragement such as "The experiment requires that you continue" and "You have no other choice—you must go on."

So, each teacher in this experiment was led to believe that not only were they shocking the learner for each wrong answer, but that the shocks were getting stronger each time. After a few shocks, the learner would play a tape of prerecorded yelps, yells, screams, and so forth. The actor would also pound on the wall and complain of a heart condition. At this point, most teachers wanted to stop the test but were told they would not be held responsible if anything went wrong. Some of the subjects started laughing nervously and showing other signs of extreme stress. The experiment was halted either after four verbal prompts from the experimenter or after the teacher had given the maximum voltage three times in a row. In the first trial, 65% of the subjects administered the maximum voltage.

The Sleep Creep

Dr. William Sargant is only one of many scientists who have experimented with deep sleep therapy, which was a psychiatric treatment in which drugs and other methods were used to induce a coma in patients suffering from mental disorders. What sets Sargant apart from the pack was his insistence that he need not get his patients' consent first.

The idea behind deep sleep therapy is that once in a coma, scientists could administer all sorts of drugs and electric shocks that would cure the patient of schizophrenia or other ailments. Not only did this not work, but at the height of the Cold War, Sargant went further, doing research on reprogramming people while they were in a coma. He developed methods for creating false memories and worked with the CIA on mind control. The CIA stopped funding the project, saying it produced only "amnesiacs and vegetables," who, thankfully for Sargant, couldn't remember never filling out a consent form.

In Case of Emergency, Fill Out Insurance Forms

It's the 1960s, and you're one of ten soldiers flying in an aircraft during a routine training mission. Suddenly, the plane lurches dramatically and begins quickly to lose altitude. The pilot gets on the intercom and explains that the plane is about to crash into the ocean. At this point, you're freaking out and kissing your sweet life goodbye. Imagine next a flight attendant passing out insurance forms that you and your fellow soldiers have to fill out before you crash. What? This is exactly what happened, but the plane wasn't in any danger of crashing. It was all a harmless science experiment designed to see how extreme stress affects a person's cognitive ability. The answer: a lot (especially since nobody informed the soldiers they were participating in an experiment of any kind). In a funny side note, the experiment was abandoned only after researchers found out that soldiers in earlier trials had written warnings on the airsick bags for the next unsuspecting subjects.

Sir, you need to fill this out before we crash.

Not Good

Can you handle one more of these harrowing stories? Well, here goes nothing. In the 1930s, Dr. Wendell Johnson, a leading speech pathologist, wanted to better understand his own stuttering. Prevailing theories at the time were that it was caused by a physical disability of some sort. Johnson believed that children were often labeled as stutterers at very young ages, and that the label itself worsened children's speech. To test his hypothesis, he had one of his graduate students select twenty-two children from an orphanage, ten of whom stuttered. The children were split evenly into two groups. One group was labeled normal speakers and the other as stutterers (no matter whether or not a child indeed had a stutter). The normal speakers received positive therapy, while the stutterers were belittled even if they weren't stuttering. This went on for four months, and many of the children in the stuttering group were traumatized for life. Often called the Monster Study, Johnson hid the experiment, fearing for his reputation, and it only became public in 2001.

OOPS

Have you ever gone searching for your car keys and inadvertently found that credit card you reported stolen or the phone number of that cute person you met at a bar—five years ago? Sometimes scientists, searching for one thing, find something else entirely.

How Sweet It Is

In at least three cases, scientists looking for answers to common maladies discovered artificial sweeteners instead.

In 1937, Michael Sveda was working on an anti-fever medication. He put his cigarette down (ah, remember the days when you could smoke in the lab!), and when he picked it up and took a drag, he discovered that he had mistakenly gotten the chemical he was working with, cyclamate, on the cigarette, which now tasted really sweet.

In 1965, a chemist named James Schlatter, while working on anti-ulcer drugs, synthesized aspartame, and mistakenly licked his contaminated finger. Yummy, he thought, and aspartame was born—a sweetener that's 200 times sweeter than sugar.

Meanwhile, while working on coal tar derivatives at Johns Hopkins University in 1878, Constantin Fahlberg accidentally discovered its sweet taste. Fahlberg applied for several patents for his new substance, which he named saccharin. There's also a similar story about the discovery of sucralose.

Explosive Results

While experimenting with urine one day (don't ask), the seventeenth-century alchemist Hennig Brand discovered phosphorus. That's good, right? Well, not for Brand, who was attempting to create the philosopher's stone, a substance that supposedly turned metals into gold and made humans immortal. He practically ignored his discovery and let others run with it.

When the Wife's Away

The nineteenth-century German chemist Christian F. Schonbein liked to experiment at home in the kitchen. His wife forbade the activity, so he could only play when his wife was away. One day, while experimenting with nitric and sulfuric acids on his wife's countertop, he spilled the mixture. The quick-thinking scientist grabbed his wife's apron and mopped up the mess. He thought he had gotten away with his secret kitchen folly when the apron spontaneously combusted. (No word on whether or not his wife was wearing it at the time.) The chemicals combined with the cotton of the apron to form what would be named *guncotton*, which would eventually be used in firearms.

Is That a Chocolate Bar in Your Pocket?

In 1945, Percy Spencer, a self-taught engineer with a sweet tooth, was testing a new vacuum tube called a magnetron for the Raytheon Corporation when he noticed that the

Damn—Helga would have killed me for this.

chocolate bar in his pocket had melted. Intrigued, he then placed some popcorn kernels near the tube. Instant popcorn! Next up, an egg, which exploded all over a curious colleague. Within a year, Spencer and an associate created a secret project called The Speedy Weenie, and the microwave oven was born. The first commercial oven was tested in 1947. It weighed 750 pounds, was nearly six feet tall, and it cost around $5,000. Needless to say, it took some tinkering to get it right.

MASTERS OF THE OBVIOUS

If we as a race of intelligent animals decided to take everything at face value, we'd still believe the world was flat and that Tiger Woods was nothing more than a devoted family man. Thankfully, we're curious and don't take things for granted. However, there are times when scientists should just accept what we all know to be the truth and leave it be. It would certainly cut down on the number of times scientists hear "Duh."

Those Lying Liars

A 2003 article in *The Observer* confirmed what everyone

already knew: Politicians lie. Glen Newey, a political scientist at the University of Strathclyde in Scotland, described a study that determined that not only do politicians lie, but that lying is an important part of modern politics. "Politicians need to be more honest about lying," he said. The main cause of lying, according to Newey, is increased probing by the public into areas that the government doesn't wish to talk about.

The Eye of the Beer Holder

Beer goggles is a phenomenon in which the consumption of alcohol supposedly transforms regular-looking people into stunning beauties . . . at least until the next morning. Does it exist? Of course it does. But thanks to the team of researchers at the University of Bristol in the United Kingdom, beer goggles is now a scientific fact. The scientists devised a controlled experiment in which eighty-four heterosexual students were randomly assigned either a nonalcoholic lime drink or one with vodka in it. They were then shown photos of men and women their own age. Both men and women who consumed the alcohol rated the faces as more attractive than did the control group. The tipsy subjects even rated people of their own sex as more attractive.

Meanwhile, researchers at Yale University have come to the conclusion that alcohol encourages us to engage in behavior we would otherwise avoid. Students in the study reported that they were more likely to engage in risky sexual behavior after drinking.

Eat More, Gain Weight

Dr. Brian Wansink, the founder of the University of Illinois' Food and Brand Lab and the author of *Mindless Eating: Why We Eat More Than We Think*, set up several food experiments that show that the more food people are given, the more they will eat—no matter if they are full or think the food tastes good. "In the obesity war, portion size is the first casualty," said Wansink. Dr. Barbara J. Rolls took on the same problem for the *American Journal of Clinical Nutrition* and found that big portions might contribute to obesity.

Things That Hurt

Psychologists at the University of California have discovered that the same part of the brain that's responsible for the response to physical pain becomes activated as a result of social rejection. Hence, not only is breaking up hard to do, but it really hurts.

A study in the October issue of the journal *Arthritis Care & Research* found that more than 60% of older women with foot pain regularly wore high heals at some point in their lives. The study found no connection between foot pain in men and the shoes they wore. The researchers concluded that it's probably because men don't wear high heals.

A study in the journal *Archives of Internal Medicine* has concluded that eating a lot of red meat is associated with a higher risk of death from all causes, including cancer and heart disease.

Scientists at the University of Glasgow in Scotland tested smokers six weeks after they had quit. The researchers found that they had improved their lung function by 15%.

Scientists at the Wake Forest School of Medicine in North Carolina concluded recently that getting drunk makes people lose their balance.

A study of 5,000 children by the University of Southern California found that living in residential areas with high traffic-related pollution significantly increases the risk of childhood asthma.

The Bikini Project

Believe it or not but showing pictures of scantily clad females to men turns women into sexual objects in these men's minds. The study, conducted at Princeton University, found that the medial pre-frontal cortex, which helps to stem a man's hostile, sexist thoughts is deactivated when shown images of women in bikinis. "It's as if they are reacting to these women as if they are not fully human," said Dr. Susan Fiske, who led the experiment.

The Car Makes the Man

Dr. Michael Dunn of the University of Wales Institute has confirmed that women find men who drive expensive cars more attractive. Dr. Dunn's team showed women pictures of the same man sitting in two different cars: a $140,000 Bentley Continental and a beat-up Ford Fiesta. The dude in the Bentley won out overwhelmingly. Conversely, men tested in the same way were not impressed by the car. When

Wow, great car. And he's looking better by the minute.

asked if this is evidence that women are shallower than men, Dr. Dunn (assuring that no woman would ever look at him again) responded, "Let's face it—there's evidence to support it."

Jock Science

In 2007, scientists at the University of Alberta studied more than 200 elementary school children in western Canada and found that kids who were perceived by others as having good athletic skills were more popular while those without these skills most likely felt isolated and unhappy.

Meanwhile, across the ocean, scientists at the Centre for Appearance Research at the University of the West in England found that adolescents get bullied because of the way they look and this hurts their self-esteem.

Rock Hard, Die Young

A study in the *Journal of Epidemiology and Community Health* called "Elvis to Eminem: Quantifying the Price of Fame through Early Mortality of European and North American Rock and Pop Stars" examined the lives of more than 1,000 musicians. What did the scientists find? A rock-

Death plays a mean lead guitar.

er's mortality rate is two to three times that of those who don't rock.

Heroes, Nerds & Psychos: The Scientists

I F we're going to delve into the people behind the science, we might as well begin with Einstein. On one hand, the value of his scientific work is inestimable. On the other, he really knew how to screw around with women's feelings and lives. He dumped his first love (though he still sent her his laundry to do) for Mileva Maric, a rising physics and mathematics student, who helped Einstein with some of his earlier works (without any acknowledgment). Their marriage ended while he was cheating on her with his first cousin Elsa. He may have also fooled around with Elsa's sister, and would have married Elsa's daughter if she had been interested (smart woman). In short, Dennis Overbye, author of *Einstein in Love*, probably summed it up best: "If he was around, I'd love to buy him a beer . . . but I don't know if I'd introduce him to my sister." The stories that follow look at the screwed-up people behind the science and the sometimes screwed-up things they believed.

ALL ABOUT EINSTEIN

Asked to name two scientists, most people say Einstein and then stop to scratch their heads a little bit. So, it can't hurt to lead off this section with the superstar of all things science, weird hairdo and all.

Einstein's Greatest Hits

One of the best things about Albert Einstein was how personable he could be (as long as he wasn't in love with you). He was extremely quotable and oftentimes silly. Here are a few of his greatest hits:

Einstein once declared his second greatest idea (after the theory of relativity) was to add an egg while cooking soup in order to produce a soft-boiled egg without having an extra pot to wash.

After Einstein fled Germany in 1932, 100 Nazi professors published a book condemning his theory of relativity. The book was creatively titled *One Hundred Authors against Einstein*. When told of this book, Einstein said, "If I were wrong, one professor would have been enough."

One time, when asked to explain the general theory of relativity, he replied, "Put your hand on a hot stove for a minute, and it seems like an hour. Sit with a pretty girl for an hour, and it seems like a minute. That's relativity!"

Not known as a particularly stellar mathematician, Einstein once remarked, "Since the mathematicians have invaded the theory of relativity, I do not understand it myself anymore."

Once when speaking at the Sorbonne in the 1930s, Einstein said, "If my relativity theory is verified, Germany will proclaim me a German and France will call me a citizen of the

Everything's relative and those are relatively nice!

world. But if my theory is proved false, France will emphasize that I am a German and Germany will say that I am a Jew."

"In the past it never occurred to me that every casual remark of mine would be snatched up and recorded. Otherwise I would have crept further into my shell."—Albert Einstein to Carl Seelig, his biographer

One of Einstein's colleagues asked him for his telephone number one day. Einstein reached for a telephone directory and looked it up. "You don't remember your own number?" the man asked, startled. "No," Einstein answered. "Why should I memorize something I can so easily get from a book?"

Einstein's Brain in a Box

Since he was such a genius, lots of people were interested in Einstein's brain. However, nobody was quite as interested as Thomas Harvey, the pathologist who performed the autopsy on Einstein's body after his death in 1955. Without anyone's consent (certainly not Einstein's, whose wish was for his whole body to be cremated), Harvey removed the brain, photographed it from many angles, and then had it dissected into tiny blocks. (He also removed Einstein's eyes and gave them to Einstein's eye doctor! They remain in a safe-deposit box in New York to this day.) Harvey was fired a few months later for refusing to give up the brain, but that isn't the end of the story.

First, he put the brain cubes into two formalin-filled mason jars, and then stored them in his basement. Harvey's marriage ended shortly thereafter, and he left Princeton to look for a job. But he had to return quickly to get the brain after his wife threatened to throw it out. Over the next twenty or so years, Harvey fielded questions about the brain, and he always said that he was about a year away from publishing the results of his studies. He also gave several pieces to different researchers. It wasn't until 1978 that journalist Steven Levy tracked down Harvey and the brain in Wichita, Kansas. (The brain was still in the jars safe inside a cardboard box.) In the early '90s, Harvey, along with a freelance writer, drove from New Jersey (where Harvey had relocated) to California to meet Einstein's granddaughter. Harvey took the brain with him and stored it in the trunk of his Buick Skylark. He even offered the brain to the granddaughter, but she didn't want it. Finally, Harvey brought the brain back to Princeton in 1996, saying "Eventually, you get tired of the responsibility of having it."

There have been several tests done on the brain cubes, and many scientists have studied the photographs. Though the results are inconclusive, scientists report that Einstein's brain weighed less than the average adult male brain, but his cerebral cortex was thinner and the density of neurons was greater. He also had an unusual pattern of grooves on both parietal lobes, which could have helped his mathematical abilities. Finally, his brain was shown to be 15% wider than average brains.

A FEW PLANETS SHORT
OF A SOLAR SYSTEM

Speaking of famous scientists, let's take a gander at the secret lives of other celebrated scientists: their personality quirks, their neuroses, and other attributes that place them closer to nuts along the normal-to-nuts continuum.

Newton's Sins

Sir Isaac Newton, perhaps one of the most influential scientists ever, spent a large part of his life attempting to turn common metals into gold and decoding the Bible. He wrote loads on religion, translated the Book of Daniel, and believed fervently that the end of times was near. First would come war and plagues, and then the second coming of Christ, followed by a thousand-year reign of saints living on Earth.

And who would be one of those saints? Newton, of course. That, despite the debtor's ledger of sins, which as a young man, he kept track of in his journal. Of his sins he thought worthy of mention, some favorites include:

1. Robbing my mother's box of plums and sugar,
2. Calling Derothy Rose a jade
3. Punching my sister
4. Wishing death and hoping it to some
5. Threatening my father and mother Smith to burn them and the house over them
6. Having unclean thoughts, words, and actions, and dreams
7. Squirting water on thy day
8. Making pies on Sunday night.

He Put the "Mad" in Mad Scientist

Nikola Tesla (1856–1943) was a mechanical and electrical engineer who today is known as one of the main contributors to the birth of commercial electricity. He formed the basis of modern alternating current, made great contributions to research into electricity and electromagnetism, invented the radio and remote control, and contributed to early robotics and nuclear physics. He was friends with Mark Twain and lifelong enemy to Thomas Edison. But, like all the other scientists in this section, Tesla was eccentric. Beyond eccentric, actually. First off, he was obsessive compulsive. Intensely scared of germs, he refused to touch anything that might have dirt on it. Next, he disliked anything round, was repulsed by jewelry (especially pearls), and obsessed with pigeons. (He would be seen walking around New York City covered in pigeons, sometimes taking some back to his hotel with him.) Also, among his amazing accomplishments lay some real whoppers, including his idea to create a teleforce particle beam weapon, which was called a peace ray, that could shoot down "10,000 enemy planes," an antigravity airship, time-travel machines, and a camera that would take pictures of images from your imagination.

Tesla was called "the man who invented the twentieth century." You can say that again.

The Dirt Doctor

Louis Pasteur (1822-1895), known as the father of modern microbiology and who invented the pasteurization process and developed several vaccines, certainly knew his germs. Perhaps that's why he was so amazingly afraid of them. His fear of dirt and disease was so great that he refused to shake hands with anyone and would sometimes produce a portable microscope at dinner parties to make sure the food being served was okay to eat.

Nuts about Beans

Pythagoras (of the theorem of the same name) wasn't just a mathematician and physicist, but also a founder of a religion. Among the religion's tenets are the belief that the soul

Wait a minute! There's a microbobe on my meatball!

is reincarnated and the view that it's bad to eat beans.

The Original Party Scientist
Tycho Brahe (1546–1601) was a renowned astronomer who

made the most accurate astronomical observations of his time. His data was used by his assistant Johannes Kepler to derive the laws of planetary motion. Brahe, the richest man in Denmark, was a bit of a hot head, and while a student, he quarreled with another student over who was better at math. This led to a duel in which Brahe lost part of his nose. He replaced it with a homemade gold and silver prosthetic. Brahe, who was rather fond of the drink, threw mammoth feasts that always included a clairvoyant dwarf who dressed as a clown and sat under the table during meals. Party animal that he was, Brahe kept a tame elk on the premises that died after drinking too much beer.

Like Father Like Son

William Buckland was a rock and mineral expert and professor of geology at Oxford in the 1820s who discovered coprolites, which are fossilized dinosaur poop. He really liked coprolites to the point where he once had a dining-room table created from them. There's something, however, he liked doing even more than collecting dino poop, and that was attempting to eat every creature in the animal kingdom. He ate mole meat, mice dipped in batter, puppy, hedgehog, crocodile, jackal, and so forth. His son, Frank, following in his father's footsteps, also had a taste for the natural world. Although trained as a surgeon, Frank devoted his life to leading the Society for the Acclimatization of Animals, which worked to bring exotic animals into the United Kingdom to use as food. The society was known for its dinners, which boasted kangaroo and more. Frank went above and beyond the call of duty, however, and once ate slices of flesh from the head of an old porpoise (boiled and fried). When alerted of the death of any animal at the London Zoo, he would dissect the animal, and eat it. Once when he learned that a leopard had been buried without his notice, he dug it up and cooked it.

Don't worry son, it'll taste just like chicken.

DISCOVERING THE FUNNY BONE

Luckily for scientists and other lab rats, researchers at Washington State University have found that nearly 40% of people will laugh at a bad joke, while fewer than one in one hundred will voice displeasure. Why? Well, Dr. Nancy Bellmade, who headed the research, doesn't know for sure, but she believes that it's funny when we're let down by humor, and that we laugh at how bad a joke can actually be. But perhaps we are being unfair to scientists. Maybe they are simply perceived to be without humor, but are in fact the life of the party. Read along and decide for yourself.

By the way, this was the bad joke 40% of the people laughed at: What did the big chimney say to the little chimney? Nothing. Chimneys can't talk. Ha, ha.

The Name Game

One of the best parts about being a scientist (besides the

awesome lab coats) is the right you have to name something you discover. Many scientists take this opportunity to flex their funny bones. Here's a small collection of some of the more humorous names scientists have come up with.

Names that rhyme: *Adonnadonna primadonna* (fossil) named after Dionne and the Belmonts, *Raffia ruffia* (plant), *Rana bwana* (South American frog), *Alouatta ululata* (howler monkey).

Named after rock stars: *Aegrotocatellus jaggeri* (a species of trilobite) and *Anomphalus jaggerius* (a fossilized mollusk—somewhat fitting!), *Myrmekiaphila neilyoungi* (a spider), *Funkotriplogynium iagobadius* (another trilobite named after James Brown: iago = James, badius = Brown), *Preseucoila imallshooupis* (a wasp).

Named after politicians: *Agathidium bushi, Agathidium*

cheneyi, and *Agathidium rumsfeldi* (all slime-mold beetles that feed on fungus). Two entomologists who named the beetles say it was out of admiration for their principles and not because the politicians looked like beetles (or vice versa).

Wordplay: *Ba humbugi* (a snail), *Chaos chaos* (a protozoan), *La cucuracha* (a moth), *Leonardo davincii* (also a moth), *Zyzzyxdonta* (a snail with characteristics opposite to those of the *Aaadonta* snail).

Names of molecules: *Arsole, megaphone, luciferase, fukalite* (gets its name from the Fuka mine in the Fuka region of southern Japan), *erotic acid, moronic acid, windowpane, constipatic acid, fucitol, diabolic acid, dogcollarane, bastardane, butanal, vaginatin.*

Now for Something Completely Different
In 2005, a new species of lemur made the news, not be-

cause of its discovery (which actually occurred in 1990 and took fifteen years to verify), but because the anthropologists who discovered the lemur named it after John Cleese, one of the creators of *Monty Python's Flying Circus* and an animal activist who promoted the plight of lemurs in his film *Fierce Creatures*. But the *avahi cleesei* isn't the first scientific discovery named after a Monty Python alum. In 1985, a fifteen-million-year-old python fossil was named *montypythonoides riversleighensis*. (It was renamed later by another scientist who realized it wasn't a new genus.) In 1997, an asteroid was named Monty Python, and in fact, there's a series of asteroids named after each member of the comedy troupe. And who said only nerds watch Monty Python? Oh, wait. . . .

Zappafish

In 1982, Ferdinando Boero, an Italian scientist, came up with a cunning plan to meet his hero, musician Frank Zappa. First, he would ask for a grant to study the taxonomy and ecology of local jellyfish fauna at the Bodega Marine Laboratory at the University of California, Berkeley. He got it. Next, he would find a new species of jellyfish and name it after Zappa. He did it (naming the creature *Phialella zappai*). Then, he would tell Zappa about it, and they would become friends. He did and they did. Zappa said, "There is nothing I'd like better than having a jellyfish named after me." He even dedicated a live show to Boero and his jellyfish. There's no word if Zappa ever performed a concert for the spider that's named after him—*Pachygnatha zappa*. Or the planet (Zappafrank). It's the reward you get for naming your kids Moon Unit, Dweezil, Ahmet Emuukha Rodan, and Diva Thin Muffin Pigeen.

Everybody Jump!

In 1976, the respected astronomer Patrick Moore stated on a BBC radio program that because of the rare astronomical occurrence of Pluto passing behind Jupiter, Earth's gravity

But it was on the 6:00 news!

would be greatly decreased at the precise moment of 9:47 A.M. And if you jumped in the air at that exact moment, it would feel like you were floating. By 9:48 A.M., hundreds of listeners called in claiming to have felt the decrease in gravity, with many describing how they briefly floated. The date of this radio broadcast? April 1.

Speaking of April Fool's Day

In 2004, nature.com reported that the increasing force of trade winds had slightly accelerated the spin of the Earth. That meant that the length of the day had decreased over the past century and we would have to lose a day once every 100 years. The website proposed that the current day, April 1, be removed from the calendar.

Pi in the Face

In 1998, the physicist Mark Boslough, frustrated by Alabama's attempts to stop teaching evolution to school children, decided to poke fun at the state's legislators by penning a fake article. Published in the April issue of *New Mexicans for Science and Reason*, Boslough's article claimed that the Alabama state legislature had voted to change the value of pi from 3.14159 to the "Biblical" value of 3. The parody made its way onto the Internet, and before you could say "Gosh, it was just a joke," Alabama had its hands full with angry phone calls.

Pulling the Bigon Over Our Eyes

On April 1, 1996, *Discover Magazine* reported that physicists had mistakenly found a new particle, which they named Bigon. Discovered by accident (when a computer connected to a vacuum-tube experiment exploded), the particle materialized only for a few millionths of a second, but it was the size of a bowling ball. Physicist Albert Manque went on to explain that he thought Bigon might be responsible for such unexplained phenomena as ball lightning, earthquakes, migraines, spontaneous human combustion, and the unexplained failures of equipment and souffles, and perhaps even the collapse of the walls of Jericho. Despite the dateline of the article and the last paragraph just about letting everyone in on the joke, the magazine received huge amounts of mail in response to the story.

The Breath of Life

Johann Heinrich Cohausen, an eighteenth-century physician, liked to write satire to see if he could hoodwink anyone, even though he would usually give the joke away in his article. One such treatise, called "Hermippus redivivus," claimed that we could prolong life by drinking an elixir made from the breath of young women.

When Penguins Fly

In 2008, the BBC announced that camera crews filming near the Antarctic captured footage of penguins flying. They ex-

This flying stuff ain't what I thought it would be.

plained that instead of enduring the harsh Antarctic winter, the penguins flew to the rainforests of South America, where they "spend the winter basking in the tropical sun." The BBC even produced video footage of the penguins taking off. It became one of the most popular videos on YouTube.

The Weather Machine

In 1981, the *Guardian* reported that scientists at Britain's research labs in Pershore had developed a machine to control the weather. The article went on to say that scientists promised that the United Kingdom would have long, warm summers with rainfall only at night, and snow for every Christmas.

Shark Attack

In 1981, the *Herald-News* in Roscommon, Michigan, reported that three lakes in Northern Michigan had been selected to host a study on the habits of freshwater sharks. Two thousand sharks were to be released into the lakes. The federal government was said to be spending $1.3 million on the research. County officials in Roscommon protested the experiment, citing the safety of local fishermen. The newspaper received lots of really angry mail over this fake, silly story.

WHAT THE HELL WERE YOU THINKING?

We turn now from scientists with senses of humor to scientists with seemingly no moral centers. Scientists who would do anything to prove their theories, whether it be traumatizing children by only speaking to them in Klingon or performing experiments on a guy with a hole in his gut (instead of, you know, closing up the hole in his gut).

Smallpox for Small Tots

Sure it was the 1700s, and yes, Edward Jenner had a very important point to prove to the world, but his methods . . . well, let's just say he should have pricked on someone his own size.

In Jenner's day, smallpox was a rampant killer. One in three who contracted the disease died, and the survivors were often disfigured. Though others had shown that exposure to the less-threatening cowpox gave immunity to smallpox, it was not well known and the idea had many naysayers. Jenner proved them all wrong and saved countless lives with a simple experiment. He extracted a small amount of cowpox from a milkmaid and injected it into a subject. Seven weeks later, he injected the same subject with smallpox. The subject didn't get sick, and the rest, as they say, is history. But wait! Who was the brave subject who risked his life for science? His name was James Phipps, and he was eight years old at the time. But, hey, give it up for Mr. Phipps, who agreed to let Jenner expose him to smallpox several times for years afterward. He never got smallpox, and Jenner became a hero instead of a child killer.

Holed Up

In 1822, a Canadian adventurer named Alexis St. Martin was accidentally shot in the stomach. The shot blew a hole the size of a fist through his side and into his stomach.

He was not expected to survive, but a U.S. Army surgeon named William Beaumont was stationed nearby and treated the wound. Amazingly, the wound began to heal, but with a hole from the outside leading right into the stomach. In fact, if the hole wasn't covered, any food eaten would fall out. Recognizing an amazing opportunity to study the little-known process of digestion, Dr. Beaumont pretty much tricked St. Martin into becoming his servant, with one of his duties being to sit still while the good doctor pushed tubes and food on strings into the hole to his stomach. Beaumont's poking and prodding led to his discovery that digestion was a chemical process and to a career as an esteemed scientist and surgeon—even though he never once considered simply operating on St. Martin to close up the hole.

Look at the Cuddly Little Rabbit . . . Bang!

John B. Watson was a psychologist who, while putting together his theories on behaviorism, decided it was time to

Little Albert didn't have an easy go of it after Watson's experiments.

move beyond animal testing and start working with . . . babies. The hypothesis: Watson believed that emotional responses could be conditioned, or learned. To prove it, he would condition a child to fear something not usually feared (a white rabbit, for instance) by associating it with something children innately fear, such as a loud noise. It was 1920, and you can probably already predict where this story's heading.

Watson chose a nine-month-old child, who he called Albert. The baby first was shown a white rat, a rabbit, a dog, burning newspaper, and other items. Albert showed no fear of these objects. Two months later, Watson put the child in a room with a white rat. As the child reached out for it, Watson struck a steel bar with a hammer. Albert cried. Albert was shown the rat once again, but without the noise. Albert cried. Watson repeated the experiment several times. After a while, Albert became afraid of all sorts of fluffy objects, including Santa Claus. After feeling like he had proven his point, Watson set out to recondition Albert to not be afraid of fluffy things, but that part of the experiment never happened. What became of Albert? Nobody really knows, but we can all guess.

Head Cases

Charles Claude Guthrie was a physiologist in the early twentieth century who was known for his work on vascular surgery. That, and for making the world's first two-headed dog. He took one dog's head and grafted it to the side of another dog's neck. The second head had some movement, but this feat was nothing compared to what the Soviet scientist Vladimir Demikhov was able to accomplish nearly fifty years later. He transplanted a puppy's head, shoulders, and lungs to the forelimb of a second dog. This puppy could drink milk and actually bit one of the researchers. Demikhov did several of these transplants, with the longest living Frankenstein dog surviving nearly a month. Not to be outdone, the Americans recruited Dr. Robert White to work with monkey heads. He

Igor, get me two *bananas!*

actually transplanted the head of one monkey to the body of another. White believed his experiments would eventually lead to human body transplants. No word on whose bodies would be used.

Beam Him Up, Scotty

In order to better understand how children learn languages, d'Armond Speers, who has a doctorate in computational linguistics and an unhealthy love for *Star Trek*, spoke only the invented language of Klingon to his newborn son. He did this for three years and even sang the kid "May the Empire Endure," which he called a Klingon lullaby. At last report, the kid was in high school, didn't speak Klingon, and was still speaking to his father.

Let's Poison Everyone!

Sidney Gottlieb was a military psychiatrist and chemist, which was a dangerous combination considering he worked with the

CIA during the Cold War. What did he do at the CIA?

- Known as the Black Sorcerer, he was in charge of the CIA's lethal poison program.

- Gottlieb thought it would be a good idea to spray Fidel Castro's shoes with a drug that would make his beard fall out. It didn't work. His plots against Castro (all unsuccessful) included poisoning his cigars, planting explosives in a conch shell to detonate while Castro was swimming, releasing LSD into a television studio where Castro was to appear, and handing Castro an explosive pen.

- Gottlieb also put his psychology degree to use as head of the MK-ULTRA project, which worked on using mind control for spying. He slipped LSD into people's drinks to see what would happen. Oh yeah, he didn't tell them.

Smoke that, Fidel!

He was trying to see if the drug would help him develop techniques that would "crush the human psyche to the point that it would admit anything."

• He tried to assassinate Patrice Lumumba of the Congo by poisoning his toothpaste. Luckily for Lumumba, he was murdered before he could be poisoned.

• After he retired, Gottlieb took up dancing, goat raising, and working with lepers in India.

The Dance of the Dead

Giovanni Aldini (1762–1834) was a physicist who liked electricity. A lot. What did he do with electricity? He electrocuted animal carcasses for audiences throughout Europe. People flocked to see dead cows dance, jaws move, eyes blink, and more.

In 1803, Aldini was given the body of a hanged criminal. He electrocuted the face, which started to twitch. The mouth and eyes opened, and when he electrocuted the body, its arms and legs started punching and kicking around. The audience clamored to have the dead body hanged again . . . just to make sure.

Al Gore's Nightmare

Thomas Midgley, Jr. was an engineer and chemist for General Motors from the 1920s to the late '30s. He held more than 100 patents and won several awards during his lifetime. But according to environmental writer J. R. McNeill, Midgley had "more impact on the atmosphere than any other single organism in Earth's history." And not in a good way. And, it wasn't one invention that gave him this reputation, but two.

The first: Midgley, while attempting to eliminate engine knock, discovered that adding tetraethyl lead to gasoline solved the problem. Yes, he is the man behind leaded gasoline. His discovery initially helped the auto and airplane industries develop more powerful engines, giving the United States an advantage during World War II. Unfortunately, lead is also poisonous, and it killed several GM employees. Instead of admitting lead was poisonous, Midgley held a press conference and poured tetraethyl over his hands and then breathed its fumes for a minute. Mission accomplished. The world kept using leaded gasoline for another several decades. Meanwhile, it took Midgley a year to recover from the demonstration.

But we're not done yet! Discovery 2: General Motors asked Midgley to come up with a new chemical to use as a refrigerant. What did he invent, but chlorinated fluorocarbons (CFCs), also known as the chemical that put that big freaking hole in the ozone layer. He died before either of his contributions was banned.

EXPERIMENTING WITH YOURSELF

The eighteenth-century scientist Henry Cavendish performed extensive experiments with electricity, but since he had no instruments in which to measure electric current (it being the 1700s and all), he did what all the scientists in this next section did: He used himself. Cavendish shocked himself and estimated the current by how much it hurt.

You'll Poke Your Eye Out

Isaac Newton was a genius. And along with his genius came the requisite amount of crazy. Known as a paranoid, unhappy loner, Newton did some interesting self-experimenting. He once (actually, more than once) inserted a hairpin into his eye socket and moved it around "betwixt my eye and the bone as near to the backside of my eye as I could," just to see what would happen. (He saw dots.) Not content that he avoided permanent damage to his eyes, he also once stared at the sun for as long as he could bear in order to determine what effect it would have on his vision. He got away with that one, too.

I'll Drink to That

In the 1980s, two Australian doctors, Barry Marshall and Robin Warren, believed they had discovered something that would revolutionize the way doctors treat ulcers. According to the doctors, ulcers weren't caused by stress, but by bacteria. No one believed them. After unsuccessfully attempting to get rats, mice, and even pigs sick with the bacteria they found, Marshall took matters into his own hands and drank some of it. Just as he had hoped, within a few days he was puking his guts out. Twenty years later, Marshall and Warren won the Nobel Prize in medicine.

One Way to Prove a Point

In 1929, a German physician named Werner Forssmann believed that a catheter could be inserted directly into the heart. If true, the applications of such a procedure included delivering medicine and measuring blood pressure. Most believed that inserting anything into a heart would kill a person, so

Forssmann did what any determined head case would do: He pushed a catheter into a vein in his arm, and threaded it more than two feet until it reached the right atrium of his heart. He then calmly walked to the X-ray department to take a few snapshots of his work. He was soon fired from the hospital where he was employed for self-experimentation, but nearly thirty years later, Forssmann won a Nobel Prize for his discovery.

Fishing for Answers

By the eighteenth century, scientists had come up with different hypotheses on how digestion happens, but had no real way of testing any of them. Did the stomach grind the food or did food simply rot in the stomach? Lazzaro Spallanzani, an Italian priest and scientist, decided to find out. Besides swallowing linen bags with bread in them (to see if the bread was digested when the bag came out the other end) and thin, wooden tubes (also with bread inside them), he came up with an ingenious way of gathering his own gastric juices. He swallowed a sponge tied to a string, and then pulled the sponge back out after it had absorbed the juices. Thankfully, he never felt comfortable trying these experiments on other humans.

Hmm . . . Tastes Like, Urgh, Cough, Yeeeaugh

The eighteenth-century Swedish chemist Carl Scheele identified oxygen (although Joseph Priestly published his findings first), tungsten, manganese, barium, chlorine, and more. As brilliant as he was, Scheele had the unfortunate habit of tasting his discoveries. And although he somehow managed to survive a dab of hydrogen cyanide, he died at the age of forty-four, most likely due to mercury poisoning.

Eternal Stupidity

Alexander Bogdanov was a Russian physician during the early twentieth century. He believed that a steady stream of blood transfusions would grant eternal youth. So, he gave himself a steady stream of blood transfusions. It improved his eyesight! It stopped his hair loss! It killed him after he gave himself a dose of infected blood!

The Human Cannonball

John Paul Stapp was concerned about the safety of pilots; however, he wasn't much concerned about his own. During the 1940s, conventional wisdom believed a person could only survive up to a certain g-force (about 18 gs). Stapp, a U.S. Air Force officer and flight surgeon, was not one to let conventional wisdom stand in the way of proving his point, so he rode a rocket car that went from 630 mph to a standstill in 1.4 seconds, proving that with the proper safety gear, a person could withstand up to 40 gs and perhaps more. This and other experiments gave him broken bones (including his back), ruptured blood vessels in his eyes, and a hernia. What's even more interesting, it seems that he and his team were the first to coin the phrase "Murphy's Law" to describe how anything that can go wrong will go wrong. Although, the fact that Stapp lived to a ripe old age, demonstrates that Murphy's Law couldn't have applied too often with him.

The Most Radioactive Man on Earth

Dr. Eric Voice was a nuclear physicist who believed all the ballyhoo over the dangers of nuclear power was media hype. But since Voice was a man of action, he volunteered to become a human guinea pig to prove it. In 1992, he and another volunteer were injected with plutonium. Then, over the next eleven years, Voice, along with eleven others, inhaled plutonium isotopes. (Voice, who lived in Scotland, needed to have his bodily wastes carted away by armored vehicle.) In 1999, the UK Atomic Energy Authority declared that all the subjects were healthy. Voice died at age eighty, and no matter what he thought about nuclear energy, his body had to be buried in a lead-lined coffin because it was considered hazardous.

Hey Ffirth, can I have a sip of your smoothie?

This Is Just Wrong

This story is so gross and wrong on so many levels there is no other suitable headline for it. In 1793, yellow fever, a virus transmitted by mosquitoes, killed nearly 10% of the population of Philadelphia. A few years later, Dr. Stubbins Ffirth studied the disease and hypothesized that it wasn't contagious. Nobody believed him, so he turned to drastic measures. First, he found patients who had yellow fever. Ffirth then collected their vomit and wiped it into open cuts he made on his own arm. He poured some on his eyes. He fried some vomit and inhaled the fumes. Then, when he didn't catch yellow fever, he celebrated with a vomit cocktail. Yes, he drank vomit. Not satisfied, he moved onto blood, spit, and urine. Ffirth never did contract yellow fever, nor did anyone think to lock him up somewhere safe. However, his hypothesis was incorrect. The samples he collected were from late-stage patients who weren't very contagious. We'd call Dr. Ffirth lucky, but then again, he drank sick people's vomit.

SIGMUND FRAUDS

Some discoveries, inventions, and experiments are just too

good to be true. Thankfully, scientists are a curious lot who like to see for themselves, which sooner or (most likely) later, leads to the truth.

The Science of the Con

Ancient alchemists believed that by combining ingredients found in nature they could turn common metals into gold and create a remedy that would cure all diseases and cause one to live forever. Sounds nice, but it's all rubbish, and by the 1600s, real scientists were separating modern chemistry from this pseudoscience. That doesn't mean alchemists haven't had their fun at gullible people's expense.

In the 1920s, Heinrich Kurschildgen believed he could pay off Germany's debts after World War I with a mysterious ray that could make anything radioactive. This ray could also split an atom and turn it into gold and cure cancer. Thousands of dollars came in from investors all over the world before Kurschildgen was arrested for the obvious con.

Long considered the source material for Mary Shelley's *Frankenstein*, Johann Conrad Dippel (1673–1734) was actually born in castle Frankenstein. But other than that fact, there isn't a lot we know about what Dippel actually did in the castle. He studied theology and philosophy, but was also an alchemist intent on discovering the elixir of life. In his castle lab, he invented Dippel's Oil, which was an awful stew made from the bones, blood, and other parts of animals. He claimed it was a universal medicine that could cure all diseases. Did it grant everlasting life? No, but for a while, it was used as an animal repellent, a fuel for camping stoves, and insecticide. Did he throw in some human body parts stolen from graves? Probably not, but nobody knows for sure.

Sharpie Science

In 1974, Dr. William Summerlin, who was working at the Memorial Sloan-Kettering Cancer Center in New York City,

This'll fool 'em!

rocked the scientific community when he announced he had transplanted tissue from one animal to another by simply keeping the tissue in a culture for a month. He showed off his results, white mice with patches of black fur that had been transplanted from black mice. However, upon further examination (a lab assistant picked up one of the mice and some of the black skin rubbed off), it was concluded that Dr. Summerlin had simply colored part of the mice's skin with a black marker. Claiming mental and physical exhaustion, Summerlin moved to rural Louisiana to spend more time with his family.

The Piltdown Scam

Amateur archeologist Charles Dawson (1864–1916) had a knack for finding important fossils, including teeth from an unknown mammal species and three new dinosaurs. Because of these early discoveries, he was made a member of the Geological Society and the Society of Antiquaries of London, all while he was still in his early twenties. And that was before his most famous discovery in 1912, the Piltdown Man, which at the time was hailed as one of the greatest archeological discoveries. The Piltdown Man was a fossilized skull found in a gravel pit in England that supposedly belonged to an early human—the missing link. The skull was big enough for a human brain, but it had the jaw of an ape. Unfortunately for science, Dawson had simply attached the jaw of an orangutan to a human skull and filed down the teeth. Now known as one of the greatest frauds in the history of science, it took forty years to uncover the hoax . . . and even longer to figure out that Dawson had most likely fabricated most, if not all, of his archeological finds.

The Fertility Freak

He was the father of sperm banks (and so much more)! A miraculous doctor who gave hope to those having trouble getting pregnant! A horrible fraud who destroyed lives! Dr. Cecil

Waddya mean? These kids look nothing like me!

Jacobson, an early pioneer in using amniocentesis to diagnose the health of the fetus, ran a fertility clinic in Virginia in the 1980s. Although he had a loyal following and some women did report getting pregnant and having babies with his procedures, most accounts are much more heart wrenching.

First, he injected women with a hormone called *human chorionic gonadotrophin*, which causes women to register positive on a pregnancy test, even though they aren't pregnant. He told these women (one of whom was a forty-nine-year-old post-menopausal woman) they were pregnant, showed them grainy ultrasounds, warned them not to see their obstetricians, and then told them at around three months that the fetus had died.

Second, the women who came to his clinic for artificial insemination using anonymous donors were actually inseminated with his own sperm. By some accounts, he fathered more than seventy-five children. He was indicted on fifty-three counts of fraud in 1991 and was sentenced to five years in prison.

AGAINST THE GRAIN

There are certain things most people, including scientists,

agree on: The world is round, evolution happened, water doesn't have memory . . . right?

Must Have Brains

Oscar Kiss Maerth, Hungarian-born author of *The Beginning Was the End* (1969), believed that intelligence can be eaten, and that our ancestors, the apes, fed on each others' brains, which made them smarter and smarter until they became us. Unfortunately, Maerth offered no evidence for his theory, so we just have to take his word for it.

A Tiny Bit Crazy

According to Chonosuke Okamura, a Japanese paleontologist, all living things evolved from extinct mini-creatures, each of which is less than $\frac{1}{100}$ inch in length. Okamura claimed to have discovered fossils of microscopic dinosaurs, horses, princesses, and even dragons that look exactly like modern animals, except for their diminutive stature. Scientists have dismissed his findings as "blobs of mineral veins and such."

The Intelligence of Water

Jacques Benveniste was a noted immunologist back in the 1980s, when he published a report in *Nature* that stated water retains the memory of medicines that have been dissolved in it—even when the water is diluted so that none of the original medicine remains. Two crazy aspects to this story: Benveniste believed water had memory, and *Nature* published his article. But the crazy doesn't stop there. Benveniste went on to claim that the water's memory could be digitized and transmitted over the telephone. In other words, you could place a glass of water on the phone and turn it into medicine. Supposedly Benveniste is working on how to transmit over the Internet, too.

RIVALRIES

Sir Isaac Newton, in what has been described as a rare moment of humility, once said, "If I have seen further [than others], it is by standing on the shoulders of giants." What most don't realize is that Newton was actually making fun of a critic of his work, Robert Hooke, who wasn't the tallest of people. Yes, following the lives of scientists is sometimes like an episode of *Gossip Girl*.

The Bone Wars

Rarely have two scientists hated each other more than Edward Drinker Cope and Othniel Charles Marsh. These two paleontologists were part of the fossil-collecting craze that hit U.S. museums during the mid to late 1800s. Cope was self-taught, prolific, and prone to anger. Marsh came from money, published infrequently, and preferred the academic life to actual digging. They started out as friends, collecting fossils together, but things began turning sour when Cope found out that Marsh was paying some of Cope's diggers to send any fossils they found to Marsh. Things didn't get any better when in 1870 Marsh gleefully pointed out that Cope had put one of his dinosaur's heads on the wrong end. Cope rolled up his sleeves and said, "Are you ready to rumble?" And what a rumble it was.

Over the next twenty-plus years, Cope and Marsh slandered each other in print, rushed to locations to "claim" them, dynamited their secret sites as they were leaving to keep the other's team from digging, smashed fossils they couldn't take with them at sites, hired spies, and stole each other's fossils. Marsh even named one of his dinosaurs *mosasaurus copeanus*. Cope followed suit with his *anisonchus cophater*, which he named for all the Cope haters out there. Their rivalry lasted the rest of their lives, with both of them dying nearly penniless. But even after Cope died in 1897, he offered up one last challenge to Marsh. He directed in his will that his brain be measured against Marsh's to see whose was bigger.

This rivalry produced not only laughs, but also an amazing amount of information about dinosaurs. Before Cope and Marsh, there were only nine known dinosaur species.

By the time they were done, there were more than 200 new species to study. Together, they defined the science of paleontology and sparked the imagination of every six-year-old boy since.

Punk'd!

Imagine two scientists in the early 1700s pondering what to do about an arrogant colleague.

"He despises us. I want to take him down a notch!"

"Me, too! Hmm . . . what to do, what to do?"

"I know! Let's punk his sorry ass!"

And so they did. Johann Beringer was a respected professor of medicine at the University of Wurzberg in Germany. Two colleagues of his, J. Ignatz Roderick and Johann von Eckhardt, didn't like the way Beringer treated them, so they devised a plan. They knew Beringer was interested in fossils, so they hired masons to craft fake fossils out of limestone and place them where Beringer and/or his supplier would find them. But these weren't ordinary fossils; they were chiseled in the shapes of lizards, mating frogs, bird and eggs, spiders with webs, and even fossils with ancient languages on them. In fact, the fossils got crazier and crazier (one supposedly even had Beringer's name engraved on it), but Beringer bought it all. He even wrote a book about his finds along with his theories, which included that the fossils could have grown spontaneously or that they were "capricious fabrications of God."

After a good laugh or three, the hoaxers tried to tell Beringer that the fossils were fake (without letting on how they knew this), but word finally got out. Beringer sued them, and by the end, all three were ruined. Although no one knows for sure if this is true, it is said Beringer went broke attempting to purchase up all the copies of his book.

AC vs. DC

Thomas Edison believed direct current (DC) was the way

to distribute electricity. (Hey, he liked the patent royalties!) Nikola Tesla, who was working for Edison's rival George Westinghouse, believed in alternating current (AC). The Battle of the Currents that emerged in the late 1880s had little to do with which was best, but everything to do with winning at all costs.

First, it should be noted that Tesla had worked for Edison at one time, and by many accounts, Edison owed him money. Anyway, Edison, never one to sit back when money was at stake, set out to prove to the nation that AC was dangerous. Sure, he lobbied politicians, but he also spread lies about AC, and sponsored demonstrations showing AC electricity killing cats, dogs, and even Topsy, a circus elephant (who was going to be terminated anyway for killing a man). Edison liked to call the act of electrocution as being Westinghoused. After all was said and done, nothing Edison did could stop AC from taking over the world. Perhaps he should have simply paid Tesla his back wages.

Science Makes the World Go 'Round

.

SCIENCE cures diseases, makes life easier, and discovers amazing things about our world. Without science in our lives, we wouldn't know that by freezing your big toe you could cure your cold, or which car you should buy when trying to attract a mate. From ancient times to today, science has always been in the pursuit of helping, even when it doesn't.

CURE-NONES

History is full of fun "cures" that actually caused more pain than curing. From bloodletting to radioactive water and even cigarettes, thank goodness we live in an age of reason where we know that magnets cure everything.

A Blow to the Brain

Got a devil knocking around in your dome? The ancient Egyptians, Africans, and Europeans all believed that drilling a hole in your skull until you hit brain matter could

*Doctor, make the hole a bit bigger so he can
wear the leftovers as a necklace . . . if he lives.*

cure you from epilepsy, mental illness, and more. This was called *trepanation*, and doctors drilled to release the demons trapped in your head. In some cultures, if you survived the treatment, you got to wear the removed piece of skull as a good-luck charm!

Heavy Metal Madness

If you have indigestion, syphilis, a bone fracture, or if you're simply feeling a little down and out, have we the cure for you! Mercury! This highly poisonous metal was used for hundreds of years as medicine, even though its most likely side effect was death.

Haircuts and Bloodlettings

Looking for something to rid the body of those bothersome imbalances and bad humors? In medieval times, you could go to your hairstylist for a cure-all . . . that actually cured none. While trimming your locks, you could ask your barber-surgeon to also cut your flesh to get rid of some of that unhealthy blood. In fact, bloodletting was also used to treat excessive bleeding. And if you didn't like being cut, you could always opt for the leeches.

But forget medieval times. This practice was in use for nearly 2,000 years, and only fell out of favor in the nineteenth century.

More Doctors Smoke Camels

Suffering from asthma, shortness of breath, or even cancer? If you lived in sixteenth-century England, one cure offered to you would be a nice, healthy cigarette. Laugh all you want at those silly people from 500 years ago, but the "health" benefits of cigarettes were being touted by doctors through the 1950s. Cigarette ads boasting of their purity ran in professional medical journals from the 1930s through the '50s. The most infamous campaign stated, "More doctors smoke Camels than any other cigarette." The ads went on to say,

"That marvelous Camel mildness means just as much to his throat as to yours."

Side Effects

In the April 13, 1998 edition of *The New Yorker,* Steve Martin wrote a short piece called "Side Effects" that poked fun at those long and often unintentionally funny side-effect messages drug companies need to include in all their advertisements. The article begins, "Take two tablets every six hours for joint pain. Side effects: This drug may cause joint pain. . . ." Now, medicine is a great thing, and if your doctor prescribes something, you should probably take it. This, however, isn't going to stop us from poking fun at some real side effects that prove that sometimes the cure is worse than the disease.

- Mirapex was a drug developed in 1997 to control the symptoms of Parkinson's disease, but it's also used for people with restless leg syndrome. This drug, along with others that have similar ingredients, have many side effects, including hallucinations, falling asleep during the day without warning, uncontrollable muscle contractions, problems with impulse control (including gambling), loss of appetite, binge eating, weight loss, weight gain, and, to top it all off, amnesia.

- Acne can be a serious problem, and for the most severe cases, there's Accutane. And though you wouldn't suffer from *all* of these side effects, even just one of them would ruin your day: rectal bleeding, bone fractures, hepatitis, depression, and an overabundance of hair, to name but a few.

- Men who suffer male pattern baldness are sometimes given Propecia, a drug originally developed to treat an enlarged prostate condition. Side effects include de-

You're kidding—it's not the hair?

creased libido, impotence, and in a small percentage of men, breast tenderness or enlargement, and breasts that may lactate.

- There are several drugs on the market to help you stop smoking. Some patients taking Chantix have reported, "changes in behavior, agitation, depressed mood, suicidal thoughts or actions when attempting to quit smoking."

- EvaMist was created to treat menopause symptoms. Side effects include headache, back pain, and less commonly cancer, stroke, dementia, heart attacks, and blood clots.

- It's one thing to have a dry mouth after taking some medication. Perhaps it's also okay to have some vision problems, depending on the disease you're treating. But how about taking your daily dose of Vasotec (to treat high blood pressure and congestive heart failure) and have the possibility of losing your sense of smell and taste, and having blurry vision as well as ringing in your ears? Good thing your sense of touch is unaffected. You could feel around until you found your bed.

The Placebo Effect

If you're worried about side effects, read this. Voltaire once stated that "The art of medicine consists of amusing the patient while Nature cures the disease." Medicine as an art seems to be supported by the findings of Fabrizio Benedetti of the University of Turin. While working on a painkiller trial, he noticed, without surprise, that the painkiller performed better than the placebo used in the control group. However, when he gave the drug to volunteers who didn't know what the drug did, it had no effect. So, he concluded that the drug in combination with patient expectancy stimulated the production of endorphins to kill the pain. In another study, Benedetti found evidence that a placebo can influence a patient biochemically. He also reported that placebos trigger dopamine in patients, and that neurons in the brain responded to a salt solution in the same way as they did to an actual drug.

SCIENTIFICALLY UNPROVEN

Everyone knows that if you cross your eyes there's that small chance that they'll get stuck there, right? Well, no matter what your mother told you, you can't make your eyes cross permanently. Sometimes, no matter what science tells us, we can't help but believe otherwise.

Sugar High

Have you ever seen a bunch of kids after eating cake and ice cream at a birthday party? They're absolutely bonkers. Common wisdom dictates that feeding children sugar will make them hyper. I mean, we've all witnessed it. Well, feeding sugar to kids is a bad idea for health reasons, but according to scientists, sugar does not make children hyper. In fact, more than a dozen different experiments show that children who consume large amounts of sugar are no more hyperactive than those who don't. It doesn't affect children with attention-deficit/hyperactivity disorder or children with sensitivities to sugar. Researchers even go on to say that the behavior parents notice after sugar intake is in their own heads. One study even showed that when parents thought their kids were given a sugared drink (even though the drink was sugar-free), they rated the kids' behavior as more hyperactive.

Drink Up!

Drink eight eight-ounce glasses of water a day. Doctors say it. Personal trainers live by it. It's in all the literature. But is it true? First of all, nobody with any certainty can say where the 8 x 8 idea came from. It became a popular mantra in the 1980s, but according to some scientists, it's a misunderstanding of normal physiological requirements. Next, in 2002, the *American Journal of Epidemiology* published an article by a Dartmouth University scientist that found no scientific studies to support drinking eight glasses of water a day. The article concluded that you don't need to drink large quantities of water to be healthy, and that you should follow one golden rule for when to drink water: Drink when you're thirsty. Furthermore, studies also show that you can get most of the fluid you need through your food.

Brain Power

Over and over again, we hear that we only use 10% of our brains. This belief has existed for more than 100 years despite all the evidence to the contrary. Brain imaging studies show that all areas of the brain are active, with different tasks assigned to different parts of the brain. Scientists have looked for the lazy 90% of our brains and have been unable to locate it.

Post-Mortem Manicure

Do we need to have our nails and hair trimmed after we die? Despite stories that describe fingernail and hair growth in dead bodies, this is pure myth. Forensic anthropologist William Maples said, "It is a powerful, disturbing image, but it is pure moonshine. No such thing occurs." So, why the story? Supposedly, the dehydration of the body after death can lead the skin around the hair or nails to retract, making it look like they're growing.

Thanksgiving Torpor

It seems that the story of pilgrims actually eating turkey on Thanksgiving isn't the only myth surrounding this holiday. When Uncle Manny falls asleep at the dinner table after his third helping of turkey, he can no longer blame the turkey. Yes, turkey contains tryptophan, which is involved in sleep and mood control in our bodies (in fact, it is often marketed as a sleep aid). However, turkey does not contain enough tryptophan to cause anyone to fall asleep after a meal. (Turkey contains about the same amount as chicken, and pork and cheese contain more.) Basically, any large protein- or carbohydrate-laden meal can induce sleepiness because of

So, if it isn't the turkey, it must be my team.

the decrease in blood flow and oxygenation to the brain. All those Thanksgiving high balls and the boring football games also contribute.

Cold Causes Colds

For something as common as the cold, you'd think we'd have our facts straight about what causes it and what, if anything, cures it. Unfortunately, cold myths are as common as colds. Scientists have known since the 1950s that being cold doesn't give you a cold. In one study, prison inmates were kept very cold for a certain amount of time, and they caught colds at the same rate as the control group. What about if you're cold and wet? Doesn't matter. So why do people get more colds during the winter? First, people are indoors more and in closer contact with people who might already have colds. Also, cold weather may make the inside lining of the nose dry out, which can possibly lead to a viral infection. As for cures, do antibiotics work? No, they kill bacteria not viruses. Vitamin

C? No. Echinacea? No. Chicken soup? Perhaps. One study found that chicken soup may provide some relief from cold symptoms because it's a warm liquid that may help loosen thickened secretions, and its ingredients may provide some relief (as would any good selection of vegetables along with any protein that has the amino acid cysteine, which has been known to thin mucus in the lungs).

WHAT'S IN A NAME?

You think you're naming your child Borealis because it's interesting and unusual. Science has another reason. Your wife has the same last initial as you. Coincidence? Think again. You believe you are completely in control of your destiny. Nope, it was all decided for you the day your parents picked your name.

Initial Concerns

A 1999 study conducted by the University of California at

San Diego suggested that a person's monogram can affect his or her life expectancy. This Theory of Deadly Initials, as the study became known, found that people whose initials spelled words with positive connotations such as W.I.N. lived four and a half years longer than people with initials that didn't spell anything. And what about poor Penelope Ingrid Garrott and Anthony Steven Smith? According to the study, they will die nearly three years before the control group. The good news for them is that the study was disproved in 2005.

Moxie CrimeFighter

In 2005, comedian and magician Penn Jillette, following the lead of several other celebrities, chose an unusual name for his newborn daughter. He named her Moxie CrimeFighter (most likely so she wouldn't feel left out during play dates with Jason Lee's son Pilot Inspektor and Gwyneth Paltrow's daughter Apple). Science must be able to answer why celebrities choose strange names, right?

Thankfully, Alex Williams of the *New York Times* set out to find the answers in a 2006 article. Jenn Berman, a clinical psychologist in Beverly Hills, claims it's unconscious, but that celebrities feel that as a sort of American aristocracy, these exotic names function as the equivalent of a royal title. Meanwhile, Stuart Fischoff, a psychologist who has worked with Hollywood clients, believes this trend shows not only that the stars need to be different but also a fear and embarrassment that people might think of them as normal. Finally, Robert R. Butterworth, a clinical psychologist in Los Angeles, says, "The child is a part of them, not an individual. It's an appendage." No matter what, the strange names certainly give the children's psychologists a starting point.

More on Moxie

You're not famous, but you still think it would be cool for your kid to have a name like Zeppelin or Thoreau. Think

Mom always loved you better.

again. A 2009 study conducted by Shippensburg University professor David Kalist of Pennsylvania reports that the more unpopular, uncommon, or feminine a boy's first name, the greater the chance he'll end up in jail. The study goes on to say that the ridicule that often accompanies an unusual name doesn't help matters. The study even lists the top ten bad boy names in America: Alec, Ernest, Garland, Ivan, Kareem, Luke, Malcolm, Preson, Tyrell, and Walter. So, while your boring son Bob pays his taxes on time, his younger brother Ivan may truly be terrible.

Dennis Smiley, DDS

Dr. Brett Pelham, a professor of psychology at the University at Buffalo, has come up with a theory called implicit egotism, which states that people tend to prefer people, places, or things that remind them of themselves. Some of his examples include people named Louis are likely to live in St. Louis; Dennis and Denise are both likely to be dentists; and people with last names with the same letter are likely to marry.

LOVE AND MARRIAGE

Science has long been interested in how we attract each other. What kind of clothing should we wear? What should we say, do, or buy? Science has also studied what happens next. This section takes a look at how science helps us fall in love, stay in love, avoid love at all costs, and more.

The Feet Tell the Story

How does a man tell if a woman is interested? Is it a slight blush or the bat of the eyelids? Top British psychologist Geoff Beattie of the University of Manchester believes it's the woman's feet movements that reveal her intentions and feelings. Here are the findings:

- If a woman moves her feet away while giggling in order to adopt a more open-legged stance, she's into you.

163

Her eyes were saying yes, but her feet were saying no.

- If she crosses her legs or tucks them under her body, it will be an early night for you.

- She might be feeling nervous if her feet grow still.

- If she moves her feet a lot, she may be shy. If they're still, she likes to be in control or is arrogant.

Beattie, who conducted the research for a shoemaking company (conflict of interest?), concludes that "The secret language of feet can reveal a great deal about our personality, what we think of the person we're talking to, and even our emotional and psychological state. They are a fascinating channel of nonverbal communication."

The 40% Allure

Psychologist Colin Hendrie and three female researchers discreetly observed women at a big nightclub and concluded that women who covered only 40% of their bodies attracted the most men—twice as many as women who covered up. The study said that most of the 60% of exposed body parts included arms and legs, and that women who exposed more than 40% turned men off. The most popular women combined the 40% rule with tight clothing and provocative dancing.

Kissing Cousins

According to a thirty-year study conducted in western Australia, most babies born to first cousins are just as healthy as children born from unrelated parents. Alan Bittles, an adjunct professor at the Centre for Comparative Genomics at Murdoch University, found that only 1.2% of births had a higher mortality rate, and that first-cousin children are less than 3% more likely to have genetic deformities.

Like a Virgin

In Japan, sexually active women about to get married can hire a plastic surgeon to create a new hymen. The procedure, called hymenoplasty, costs less than $200. And a Chinese company has begun distributing the Artificial Virginity Hymen Kit to conservative Muslim countries.

FOR THE KIDS

Science is always working toward making life better for us and our kids. That's just the reasoning a public health official in England must have used when he suggested to researchers that they develop a pill that would postpone puberty until children completed college. As far as we know, no scientist has taken the "puberty-postponed" challenge, but scientists are always on the lookout for ways to improve children's lives, questioning just about every one of our parenting decisions in the process.

Peer Pressure Peril

We all know about peer pressure. All those after-school specials have made us all too aware that those "friends" our parents didn't like us hanging out with could influence us to start drinking, smoking, doing drugs, stealing comic books, and holding up convenience stores. Unfortunately for the teenage entertainment industry, that's not what Joe Allen of the University of Virginia found.

Allen studied a group of middle school kids and followed them as they grew up. He found that the children exposed to peer pressure around the ages of twelve and thirteen turned out to be more well adjusted than the kids who weren't. The need to fit in as a teenager turns into a willingness to accommodate in adult relationships.

And what about those kids who learned to just say "No"? According to the study, they didn't become the independent adults the movies say they'll turn into. Within five years, they had lower grades, fewer friends, and were less engaged than the other kids in the study.

Role Models We Can Do Without

Professional athletes are role models, whether we like it or not. Hence, student athletes become role models for our children, whether we like it or not. According to the Josephson Institute (a right-leaning ethics center), in a 2007 study of more than 5,000 students, student athletes are more dishonest than their non-playing peers. Children involved in sports are more likely to cheat on tests, they learn how to cut corners from their coaches, and more often bully to get people to do what they want.

Early to School, Early to Fail

Sending your child to school before the age of six in order to give him or her a leg up on the competition may do more harm than good. A study by the National Foundation for Educational Research in the United Kingdom has found that children sent to school before mastering basic skills are more likely to drop out of college—if they make it that far. They are also more likely to suffer from low self-esteem and anxiety attacks.

Praise Doesn't Pay

You want your kids to do well in life. You worry about their self-esteem, so you praise them. Good idea? A 2007 study from Columbia and Stanford Universities found that over-praising children can lead to a belief in the children that smarts and talent are inborn and that there's nothing you can do to improve. So, by praising supposed innate ability over hard work, children end up avoiding challenging situations or activities that involve effort. This phenomenon is called learned helplessness, and it can lead to losing interest in school, falling grades, dropping out of school, and, of course, issues of low self-esteem and helplessness.

FOOD AND DIET

How did the cavemen and women eat without microwave popcorn, fast-food burgers, and beef jerky? (Okay, they had beef jerky, but not the kind you get at the gas station.) Long, long ago, our ancestors ate a naturally healthy diet and never had to worry about being overweight. (Some of that was due to having to run away from just about anything with sharp

170

teeth.) Today, however, we have messed with our diet to the point where our bodies don't always know what to do with the stuff we put in them. Enter science.

Weight Loss for the Strong-of-Heart

In a 2009 episode of her *Tyra Banks Show*, Banks introduced her audience to a dangerous and illegal century-old weight-loss tactic. It involved ingesting tapeworms.

How does it work? A tapeworm secretes proteins in the intestinal tract that makes digestion more difficult, so the tapeworm can feast on what you're eating. That means you can eat more! The only downside is that you have a disgusting worm in your guts. A disgusting worm in your guts that can grow up to thirty-five feet long! Scientists estimate that you can lose up to two pounds a week with the tapeworm diet without changing what you eat.

But here's a question. How do you remove the tapeworm? Perhaps Banks and her viewers should not only turn to the

You know, maybe I should give Weight Watchers another shot.

past for their diet, but also for the diet's eviction.

In the 1870s, Dr. Dowler removed a 135-foot-long tapeworm from a patient by having the patient eat lots of tree bark. Dr. Alpheus Meyers, however, developed a trap that consisted of a tiny metal cylinder tied to a piece of string. He baited it with food and the patient would swallow it. Once the worm poked its head into the trap, it would get caught and the good doctor could pull it out through the patient's mouth.

Fad-dy Diets

Unlike most of us, the diet industry is in great shape. Even though most scientists say the easiest way to lose weight is to expend more calories than you consume, there are hundreds of fad diets you can choose from. Here are some of the silliest.

- The grapefruit diet, which is based on the idea that an enzyme in the juice breaks down fat. Scientists say this enzyme is most likely broken down before being able to attack body fat. But you might lose weight from eating only grapefruits.

- The chewing movement espouses chewing your food thirty-two times before swallowing to aid in digestion. There is currently no evidence that this works, but you might lose weight from utter boredom at the dinner table.

- The cabbage diet includes eating a lot of cabbage, which guarantees you'll lose weight simply through having to eat all that cabbage.

- According to researchers at Vanderbilt University, fifteen minutes of laughter burns up to forty calories.

- The caveman diet involves only eating food that can be hunted and gathered. This includes lean meat, fish, veggies, fruit, roots, nuts, and bugs. No Ring Dings or coffee.

- The beer and ice cream diet tried to utilize the scientific law of thermogenesis to prove you can eat just about anything you want as long as it's cold enough. The thinking was that by consuming cold foods, your body would have to work hard to warm up the meals before you could digest them. Does it work? Not at all.

Weight Loss through the Nose

Tired of working out and eating less to lose weight? Why not try getting people to perceive that you're slimmer? That's the question Dr. Alan Hirsch of the Smell and Taste Treatment and Research Foundation in Chicago, Illinois, set out to answer. He and his team asked a 5'9", 245 pound woman to show up for each day of the experiment wearing the same clothing and no perfume. The team then gathered nearly

200 males, who were divided into groups and whose job it was to guess the woman's weight. However, each time a group entered the room with the woman, she was wearing a different scent: citrus floral, sweet pea mixed with lily of the valley, or floral and spices. The results? The men who smelled the floral and spices reduced their perception of the woman's weight by more than four pounds. And, of the men who found the floral and spice scent to be pleasant, they perceived the woman to be twelve pounds lighter.

Drink Water. It Makes You Happy.

Small amounts of lithium, a naturally occurring drug used in the treatment of bi-polar disorder, has been found in the water supply of El Paso, Texas. Dr. Earl Lawson, a biochemist from the University of Texas, says there's enough natural lithium to keep the entire city happy, and it's one reason for the city's low incidence of mental illness. Studies in Japan have found other communities with lithium in their water, and these communities have fewer suicides. The Japanese scientists concluded that purposely adding lithium to the world's water supplies could "potentially offer an easy, cheap, and substantial strategy for worldwide suicide prevention."

Lab Meat

Like meat, but having ethical concerns about eating it? Worry no more! In 2009, researchers in the Netherlands created the first real meat that didn't come from a dead animal. The meat was grown in the laboratory and no animals were hurt or killed in making it. Vegetarian groups have stated that they have no ethical objections, but the soggy piece of pork the scientists produced is a long way from being presentable, let alone edible. They believe lab meat could be on sale within five years.

Low-Fat Cows

As reported in *Chemistry & Industry* magazine in 2001, re-

searchers in New Zealand have found a natural gene muta-
tion that made one cow produce great tasting low-fat milk.
Marge, as the cow became known, makes milk with 1% fat
instead of the normal 3.5%. Needless to say, Marge is being asked to produce a lot of offspring, and ViaLactia, the
company that owns Marge, expects the first cow-produced
low-fat milk to hit the market sometime in 2011.

Naked Chickens

Avigndor Cahaner, an Israeli geneticist at the Hebrew University of Jerusalem, has created a featherless chicken. He says
that the naked chicken could be the future of mass poultry
farming in warmer countries. The new chicken is lower in
calories, faster growing, and gross to look at. Skeptics worry
about parasites, mosquitoes, and sunburn.

Spig or Pigach?

In 2002, researchers at Kinki University in Japan reported

I should have taken the down coat today.

that they spliced spinach genes into pig DNA, creating the first-ever mammal/plant hybrid. The new aniplant should be more healthful than normal pork. What's next? Chickpeas? How about some asparigoat?

MAKING LIFE EASIER

What is the ultimate goal of science? Is it to understand the world a little bit better? Prove or disprove the existence of God? Or is it more about improving our lot in life while proving just how helpless we can be sometimes?

Rolling Up Your Sleeves

Fashion-conscious scientists working for Italian fashion house Corpo Nove have developed Oricalco, a shirt that rolls up its sleeves when you get warm. The fabric of the shirt is woven from nitinol, which is an alloy that returns to its original shape when heated to a certain temperature. In other words, the sleeves shorten when it gets hotter out. So far the shirt is only a prototype, and it costs nearly $5,000. But, hey, it never needs ironing.

Techno Pints

Scientists at the Mitsubishi Electric Research Laboratories

have come up with a way for wait staff at restaurants to fill up your pint before you even think to ask for another round. They have invented iGlassware, which are drinking glasses that detect fluid levels. Coils in your table power your glass, which then detects how much beer you've got left through microprocessors in the glass.

Bright Ideas

Ever since scientists have identified the gene that makes jellyfish, fireflies, and mushrooms bioluminescent, they've been trying to figure out what to make glow in the dark. So far, we have glow-in-the-dark cats, dogs, mice, furniture, and now, best of all, the first self-illuminated Christmas tree. In 2007, Edward Quinto of the International Society for Bioluminescence and Chemiluminescence produced a glowing Christmas tree using no electricity—just the genes from a squid. Researchers are now looking into turning trees into streetlamps.

Pet Smart

There's nothing quite like having a pet. But sometimes they can be a nuisance. Thankfully, in today's everything-must-fit-my-lifestyle world, science is helping pets fit in by genetically engineering them. In 2006, Lifestyle Pets, which specializes in genetically altered pets, created the Ashera GD, which is a cat genetically engineered to be hypoallergenic. For only $27,000, you can own a cat that doesn't make you sneeze. In 2008, they produced their first gm dog.

One Smart Egg

Scientists in England have designed eggs that tell you when they (the eggs) are cooked. In order to quell national dissent over how to best boil an egg, the British Egg Information Service (seriously) has designed eggs with an invisible, heat-sensitive, ink drawing that becomes visible the moment the egg reaches the correct temperature. The eggs come in three varieties: hard-boiled, medium, or soft.

What Music Should You Listen To?

Having trouble filling up your iPod. Science can help. According to Steven Stack and Jim Gundlach (of Wayne State University and Auburn University, respectively), you may not want to invest too heavily in country music. They did a study of forty-nine metropolitan areas and found that the greater the airtime devoted to country music, the higher the suicide rate among Caucasians. Their paper hypothesizes that country music's preoccupation with marital discord, alcohol abuse, and alienation from work nurture a suicidal mood.

Meanwhile, a high school student in Suffolk, Virginia, won the state science fair with an experiment that should keep you away from the hard-rock section on iTunes. David Merrill divided a group of seventy-two mice into three groups. One group listened to Mozart for ten hours a day, another group listened to hard rock for the same amount of time, and a third group listened to nothing. Then, after listening

Hey Slash, which way to the cheese?

to the music for a number of days, he tested how the mice did in completing a maze. The first time Merrill tried this experiment, he had to cut it short because all the hard-rock mice killed each other. The following year, he tried again, and the Mozart mice cut their maze-completion time by more than eight minutes. The mice that listened to nothing cut five minutes from their time. The hard-rock mice? They added more than twenty minutes to their time.

If that isn't enough to steer you away from heavy metal, Dorothy Retallack in her 1973 book, *The Sound of Music and Plants*, told of how she played different types of music for plants. The plants that listened to jazz actually bent toward the radio. The plants that listened to country showed no reaction. They liked Bach and North Indian sitar and tabla music. And the rock and roll plants that listened to Jimi Hendrix and Led Zeppelin? They grew very tall, but were droopy with faded blooms. Most died within two weeks.

Science in the Workplace

In 1969, a Canadian psychologist named Laurence Peter created the Peter Principle, which states that people in a workplace tend to get promoted until they reach their "level of incompetence." Peter says this happens because bosses assume that people who are good at their jobs will be good at the jobs higher up on the corporate ladder. Unfortunately, that doesn't always turn out to be the case. The promoted employees have a good chance of being really awful at the new jobs. Thankfully, three Italian scientists discovered that you can overcome the Peter Principle at your office by promoting people randomly. By using computer models, they first followed the Peter Principle, and soon the virtual office was filled with incompetents. Then they reprogrammed the simulation so that people were promoted randomly, and the efficiency of the office improved.

Send in the Clowns

Having trouble getting pregnant? Shevach Friedler, a fertility specialist at the Assaf Harofeh Medical Centre in Zerifin, Israel, has a solution that will make you laugh. He was able to raise the success rate of embryo transfers leading to pregnancy from 20% to 35% by sending professional clowns in to the women's hospital rooms to make them laugh after the procedure.

Lazy Laundry

Researchers at Monash University in Australia have discovered a way to coat clothing fibers with *titanium dioxide nanocrystals* (used in sunscreen and toothpaste), which break down food and dirt when exposed to sunlight. The material has so far stood up to wine, coffee, and even ink stains.

Meanwhile, researchers are also experimenting with implanting your clothing with the *E. Coli* bacteria, which will feed on your sweat and just about anything else. On whether or not anyone would ever buy such clothing, one of the researchers responded, "We are actually crawling with microorganisms and bacteria even after a shower. . . . As long as I couldn't feel the shirt moving, I suppose I'd give it a try."

Going Green

Why not finish things off with science even making death easier (or at least greener!).

Ashes to ashes? Not very green. Dust to dust? Also, not good for the environment. If you're worried about your post-life carbon footprint, you know that cremation is out of the question. Sure it's cheep, but it sends carbon dioxide into the atmosphere. And burial requires land and embalming fluids. Perhaps you should try resomation, which is a process that liquefies your body. It uses about one-sixth of the energy of cremation, and what's left can basically go down the drain. This process was first used to dispose of mad cows in Europe, but now it's becoming a fad for the recently deceased in the three American states where it's legal.

Index